ФИЗИКО-МЕХАНИЧЕСКИЕ СВОЙСТВА ЭЛЕКТРОЛИТИЧЕСКИХ ОСАДКОВ

FIZIKO-MEKHANICHESKIE SVOISTVA ELEKTROLITICHESKIKH OSADKOV

THE MECHANICAL PROPERTIES OF ELECTROLYTIC DEPOSITS

THE MECHANICAL PROPERTIES
OF ELECTROLYTIC DEPOSITS

by

A. T. Vagramyan and Yu. S. Petrova

Authorized translation from the Russian

CONSULTANTS BUREAU
NEW YORK
1962

ISBN-13: 978-1-4684-1544-5   e-ISBN-13: 978-1-4684-1542-1
DOI: 10.1007/978-1-4684-1542-1
Library of Congress Catalog Card Number 62-8012

The Russian text was published by the
USSR Academy of Sciences Press
for the Institute of Physical Chemistry
in Moscow in 1960.

*Ашот Тигранович Ваграмян
и Юлия Степановна Петрова*
**Физико-механические свойства
электролитических осадков**

PREFACE

This monograph attempts to establish a connection between the structure and physicomechanical properties of the deposit, the rate of reduction of the metallic ions, and the adsorption and occlusion of surface-active substances and hydrogen, and presents the experimental results of the authors as well as the data of other investigators. The authors do not pretend to have given an exhaustive treatment of all aspects of this basic problem since many of the questions touched on here are under investigation in solid state physics and various fields of physical chemistry.

THE FOLLOWING SOVIET JOURNALS CITED IN THIS BOOK
ARE AVAILABLE IN COVER-TO-COVER TRANSLATION

Russian Title	English Title	Publisher
Doklady Akademii Nauk SSSR	Proceedings of the Academy of Sciences of the USSR, Section: Chemistry	Consultants Bureau
Izvestiya Akademii Nauk SSSR: Otdelenie khimicheskikh nauk	Bulletin of the Academy of Sciences of the USSR: Division of Chemical Sciences	Consultants Bureau
Stal'	Stal (In English)	Iron and Steel Institute
Uspekhi khimii	Russian Chemical Reviews	The Chemical Society (London)
Zavodskaya laboratoriya	Industrial Laboratory	Instrument Society of America
Zhurnal analiticheskoi khimii	Journal of Analytical Chemistry USSR	Consultants Bureau
Zhurnal fizicheskoi khimii	Russian Journal of Physical Chemistry	The Chemical Society (London)
Zhurnal obshchei khimii	Journal of General Chemistry USSR	Consultants Bureau
Zhurnal prikladnoi khimii	Journal of Applied Chemistry USSR	Consultants Bureau
Zhurnal tekhnicheskoi fiziki	Soviet Physics - Technical Physics	American Institute of Physics

CONTENTS

INTRODUCTION

One of the most important problems in the field of the electrode-position of metals is that of developing new heat- and wear-resistant platings and assuring their extensive application in the manufacture of various types of machine parts. The study of the physicomechanical properties of electrolytic platings is of prime importance here, and the number of papers devoted to such investigations has increased continuously in recent years.

The physicomechanical properties of an electrolytic deposit give a fundamental characterization of its qualities as a plating and determine its applicability in various technological fields. The properties of such a deposit are fixed by the presence of occlusions of foreign particles and by the degree of irreversibility of the electrode reaction, i.e., by the overvoltage. This last factor comes into play through the fact that the ion undergoing reduction acquires high energy in traversing the intense field of the electrical double layer (which is of the order of 10^7 v/cm) and then loses this energy as soon as it penetrates into the crystal lattice, thereby forming a nonequilibrium lattice with new parameters.

Special note should be made of the fact that simultaneous ionic discharge opens up very interesting perspectives for the production of platings having properties which cannot be duplicated by other means. Thus alloys which are not formed in melt crystallization may be prepared by electrolytic deposition [1]. It is known that the maximum solubility of lead in silver is approximately 1.5% at the eutectic temperature (300°) and diminishes as the temperature falls, the metals becoming almost completely insoluble at room temperature. Lainer [2] has shown that electrolytic methods can be used to prepare a lead—silver alloy containing as much as 8% lead at room temperature.

An electrolytic alloy is not in a state of equilibrium and its structure is markedly different from that of the thermally prepared alloy. It is therefore impossible to directly consider in terms of the equilibrium phase diagram an alloy which has been prepared by electrolysis since its structure and properties will both depend on the method of preparation. This can be seen merely by comparing the hardness of electrolytic deposits of such metals as chromium, platinum, and rhodium with the hardness of these same metals after annealing.

An electrolytic deposit contains metallic occlusions and large amounts of such nonmetallic substances as hydroxides, oxides, water, hydrogen, surface-active agents, and halogens [3-8], each of which affects the physicomechanical properties.

The possibility of obtaining and controlling various desirable properties has served to extend the uses of electrolytic platings in industry. Plating is widely applied to increase the wear resistance of metallic surfaces [9] (chromium, iron), to mask selected areas of steel objects during carburization (copper), for decoration (nickel, chromium, silver, gold), to increase reflective power (silver, chromium, cadmium), and, finally, to protect the base metal from corrosion (zinc, cadmium, lead, tin, nickel). No other method can, as yet, compete with the galvanic procedure for obtaining platings of high-melting metals and alloys, such as, W—Ni, W—Co, W—Cr, and Cr—Ni.

Anodic treatment is often resorted to in order to improve the mechanical properties of an electrolytic deposit and obtain a lustrous plate [10].

The necessity of studying the mechanical properties of electrolytic deposits attracted attention at the very beginning of the development of electroplating, the importance of these properties to practical applications being pointed out by Jacobi, the founder of electroplating and a member of the Russian Academy of Sciences.

The fruitful development of electroplating and the study of the characteristics of electrolytic deposits were furthered by Lentz, who showed as early as 1870 that the mechanical properties of these deposits varied markedly with the conditions of electrolysis. The fact that the brittleness of a deposit is highly dependent on the amount of occluded gas was brought out clearly in this work.

The study of the mechanical properties of electrolytic platings was advanced considerably in a paper of Miles which appeared in 1887. The fact that two types of internal stress arise in electrolytic deposits was first noted here and served to stimulate a new line of development in the electrodeposition of metals.

Among the general complex of physicomechanical properties, the elasticity, plasticity, porosity, hardness, and wear and fatigue resistance play the principal role in determining the applicability of electrically deposited metals. The presence of the base metal makes for difficulties, however, in evaluating certain of these characteristic mechanical properties of electrolytic platings and investigation is generally limited to a study of the hardness, plasticity, and porosity.

Study is also made of the internal stresses resulting from various changes in the deposit itself (orientation and dimensions of the crystals, stability of various modifications) and from alteration of the amount and condition of the occluded foreign substances. These internal stresses are sometimes of considerable magnitude, amounting, for example, to 2700-3000 kg/cm^2 [11] in the case of nickel. Internal stresses of this order must naturally have a marked influence on such properties of the deposit as porosity and fatigue resistance.

Internal stresses are almost invariably present in the electrolytic deposit and characterize its state of compression or tension. The deposit cracks under these stresses, the number of pores increases, and the corrosional protection diminishes [12]. Furthermore, these internal stresses frequently reduce the adhesion of the deposit to the base metal [13]. The presence of high internal stresses in the electrolytic platings deposited on various machine parts also reduces the fatigue resistance of the latter. There are cases, however, in which the internal stresses play a beneficent role, one of these being the preparation of the so-called porous chrome plates by electrodeposition.

The Fourth International Conference on the Electrodeposition and Finishing of Metals which was held in London in 1954 recognized the great significance of the internal stresses in metals and recommended that they be checked daily in normal bath operation [14]. The internal stress is thus an important parameter characterizing the physicomechanical properties of a plating and the present monograph, while treating other properties of this kind, will deal principally with the origins of these stresses and the various factors which affect them.

LITERATURE

1. V. I. Lainer and N. T. Kudryavtsev, Principles of Electroplating [in Russian] (Metallurgizdat, Moscow, 1953).
2. Collection, The Structure and Casting of Light Metals [in Russian] (Metallurgizdat, Moscow, 1945) p. 43.
3. K. M. Gorbunova, N. A. Shishakov, and T. I. Ivanovskaya, Zhur. Fiz. Khim. 25, 981 (1951); J. B. O'Sullivan, Trans. Faraday Soc. 26, 28 (1930).
4. J. Billiter, Principles of Electroplating (ONTI, Moscow, 1937).
5. J. B. Kushner, Metal Finish. 56, 46 (1958).
6. D. J. Macnaughton, J. Iron and Steel Inst. 119, 179 (1924); L. Guillet and J. Cournot, Compt. rend. 192, 787 (1931).
7. C. Marie and A. Buffat, J. Chem. Phys. 24, 470 (1927); E. Tesche and P. Rysselberghe, Trans. Amer. Electrochem. Soc. 59, 353 (1933); M. Schlotter, J. Korpium, and W. Burmeister, Z. Metallkunde 25, 107 (1873); W. S. Huges, Metall Ind. 33, 293 (1928).
8. G. Gardam, Disc. Faraday Soc. 1, 182 (1947).
9. Collection, The Theory and Practice of Electrolytic Chrome Plating [in Russian] (Acad. Sci. USSR Press, Moscow, 1957).
10. N. P. Fedot'ev and A. A. Khonikevich, Transactions, Lensovet Leningrad Technological Institute No. 40 (Goskhimizdat, Leningrad, 1957) p. 133.
11. Collection, Studies on the Corrosion of Metals under Stress [in Russian] (Mashgiz, Moscow, 1953) p. 86; A. L. Rotinyan and E. S. Kozich, Zhur. Priklad. Khim. 31, 424 (1958).
12. A. T. Vagramyan and Yu. S. Tsareva-Petrova, Doklady Akad. Nauk SSSR 98, 807 (1954).
13. N. P. Fedot'ev and E. G. Kruglova, Zhur. Priklad. Khim. 28, 275 (1955).
14. Galvano 23, 19 (1954).

Chapter I

CERTAIN ASPECTS OF THE ELECTROCRYSTALLIZATION OF METALS

An idea of what is involved in the electrodeposition of metals can be obtained by considering the basic principles which apply to the electrocrystallization of the simplest substances. The simplest approach is to consider the growth of the individual metallic crystals.

A hypothesis to the effect that crystal growth occurs periodically rather than continuously was formulated by Gibbs as early 1876-1878. This hypothesis has been fully confirmed by various studies [1] in which it has been shown that growth in electrocrystallization does not involve the entire expanding crystal face, but only certain sharply defined active centers, the remaining portions of the face taking no part in the electrolysis. The corners and edges of the crystal serve initially as active centers for electrolysis, and it is here that the deposition of metal and the formation of a new layer are initiated on the face. The corners and faces of the resulting steps become new active centers for further deposition of metal and the layer is thereby propagated over the entire crystal face.

Observation of the growing crystal under the microscope shows that new layer steps are being formed on its corners and edges as the layers continue to expand over the face. Crystal growth takes place only through the superposition of such layers, one on the other.

Stranski and Kossel [2] claim that crystal growth occurs predominantly on those parts of the lattice which are structurally incomplete (Fig. 1), deposition of metal in these positions minimizing the increase in the surface energy.

Experiment has shown that the depth of the expanding layer varies from case to case, with the thinnest layer showing the highest rate of propagation along the crystal face. An alteration in the current strength not only leads to a temporary change in the rate of layer propagation, but also alters the number of new layers which appear on the crystal face. These alterations disappear as soon as the change in the surface area of the crystal has brought the current density back to its original value. The rate of layer propagation varies from face to face of the crystal, being lower on faces of high surface energy than on faces whose surface energy is low. Layer propagation ceases as soon as the current is turned off, and begins once more when the circuit is again closed, provided the intervening time interval is not extensive. It is clear that the active centers are steps formed by these layers and that it is on these that the metal deposits.

The depth of the layer depends on such factors as the concentration of the electrolyte and the nature of the contaminants, and can be calculated from the following equation:

$$\rho = k_1 \frac{I}{n_S} .$$

Here ρ is the layer depth, expressed in number of atoms, I is the current density on the active surface of the growing crystal, $1/k_1$ is a constant which is equal to the strength of current required for forming a monatomic lattice plane with surface area of 1 cm^2 in 1 sec, and n_S is the number of layers formed per second on the surface in question.

In its growth on the crystal face, the layer is usually propagated in a definite direction, either along the face itself or (sometimes) spirally [3]. It has been estimated that the normal growth (without addition agents) of silver crystals from a 3 N AgNO$_3$ solution involves the deposition of layers which are approximately 60-120 atoms deep. Slowly expanding layers occasionally have a depth of several thousand atoms.

The layer depth and propagation rate vary from metal to metal. Thus Hoekstras [4] has observed that the copper layers on a copper crystal are not easily discernible and propagate with a velocity of approximately 0.00002 cm/min. The lead layers on a lead crystal are, on the other hand, clearly visible and are propagated linearly, one after the other. Even a brief interruption of the current affects the growth of these layers and they become completely irregular as a result of the marked tendency of lead to passivation.

Tin layers on a tin crystal also propagate linearly with an approximate velocity of 0.00008 cm/min. The layers on iron, nickel, and cobalt are considerably thicker, the data of Hoekstras indicating the formation of layers approximately 1000 atoms deep during electrodeposition of these metals.

Thus crystal growth does not occur uniformly and continuously over the entire surface but periodically on various active centers, with subsequent propagation over the crystal face.

Passivation can bring the growth of a face to a standstill if the rate of layer propagation is not sufficiently high. The adsorption of any kind of foreign particle on the surface can serve as a passivating factor.

The conditions leading to cessation of crystal growth have been treated by Kolschütter and Torricelli [5]. These authors have noted that a growing monocrystal of silver can become passive when the increase in surface area has reduced the true current density and rate of deposition of metal. Temporary interruption of the current leads to artificial passivation, but only of certain of the active centers. The corners, or even the entire crystal face, can be made passive by increasing the time of holding the electrode in the electrolyte with the current turned off; the whole crystal will become passive if it is held under these conditions long enough. This sequence of passivation appears natural if it is considered that the corner atoms of the crystals are most highly unsaturated and therefore have the highest adsorption capacity, while the atoms on the edges are less active and those on the faces less active still. The tendency to passivation varies over a crystal face in which the atomic packing is nonuniform.

Electrolysis not only involves a continuous alteration in the dimensions of the expanding sectors on the various faces, but there is also a nonuniform growth of the individual crystals resulting from inequalities in current distribution. The cessation of growth of "old" crystals and the appearance of "new" ones leads to a continuous redistribution of current during the formation of a polycrystalline deposit. The metallic ion content of the solution surrounding the growing crystal diminishes, i.e., the so-called concentration polarization* increases, there is a redistribution of the current flow lines and the metal begins to deposit on those portions of the cathode where the concentration is highest. An increase in the rate of depletion of the solution surrounding the growing crystal leads to a more rapid redistribution of the current flow lines and a more finely crystalline deposit is produced.

Passivation sets in when redistribution of the lines of current flow has reduced the current density on the crystal below a certain definite limit. The "old" crystals do not continue to grow when the concentration of discharging ions in the solution is once more built up but "new" crystals are formed. It is clear that the structure of the deposit will be the finer the more rapid the redistribution of the cur-

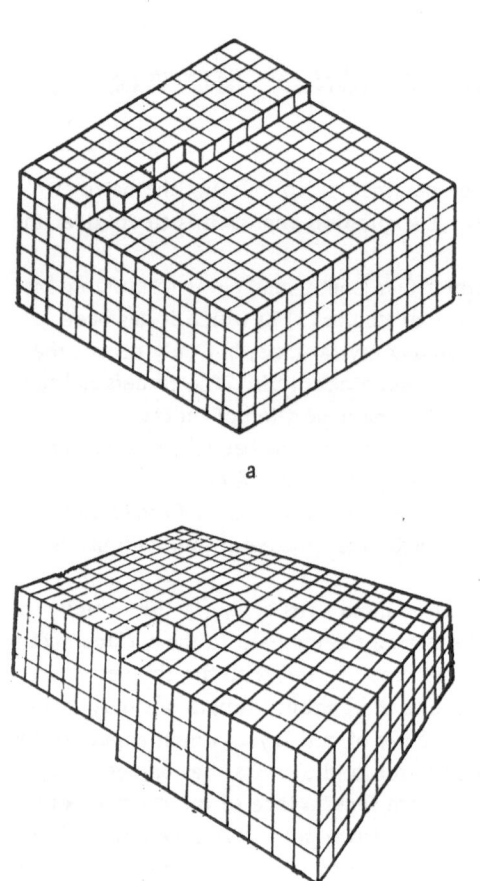

Fig. 1. Active centers on an expanding crystal face: a) Linear growth; b) spiral growth.

rent flow lines and the more frequent the passivation of the crystals. The probability of the formation of "new" crystals on the less active centers increases with an increase of the electrode polarization (Fig. 2) as does the number of crystals which are formed and the degree of fineness of the deposit.

It follows from what has just been said that electrolysis involves a competition between deposition and metal passivation.

It must be noted that passivation during electrolysis is an extremely important factor in fixing not only the structure of the electrolyte metal, but also the rate of discharge of metallic ions.

The metals can be arbitrarily divided into three groups on the basis of the difficulty of iron reduction at an electrode.

The first group contains those metals which discharge rapidly and separate out at low overvoltage; in it are tin, cadmium, zinc, copper, silver, and others. It is characteristic of the metals of this group that surface passivation

_____

* The term, "concentration polarization" is used to designate the displacement of the electrode potential resulting from an alteration in the concentration of the metallic ions in the layer surrounding the electrode during passage of the current.

occurs slowly, separation of the metal taking place principally on the active centers of the electrode without appreciable chemical polarization.* The rate of ionic discharge is proportional to the fraction of the electrode surface which is free of foreign particles. These metals can be obtained with comparative ease from solutions of their various salts since passivation is much slower than deposition. The result is that these metals can be prepared in pure form without appreciable occlusions of foreign substances.

The second group contains metals which separate out at high overvoltages; it includes iron, cobalt, nickel, chromium, and manganese, among others. The tendency to passivation is known to be pronounced in these metals. Continued discharge of such metallic ions is retarded to a considerable degree by the presence of a film of foreign particles which forms rapidly on the electrode surface. The existence of such passivating films has been repeatedly confirmed experimentally. Thus Hoekstras has shown that the discharge of nickel ions on a nickel electrode is quite rapid when the surface film is continually scraped away. It is probable that these metals cannot maintain clean surfaces for any extended time, adsorption of foreign particles giving rise to passivation and passage into a more stable state. Because of the high rate of passivation, deposition of these metals can take place only from certain salt solutions and only under certain definite conditions of electrolysis. Moreover, the deposit obtained is not of high purity being contaminated by a certain amount of oxides, hydroxides, hydrogen, etc.[6].

The third group contains those substances which have not as yet been obtained in metallic form from aqueous solutions [7]. Included in it are molybdenum, tungsten, uranium, niobium, titanium, and tantalum. The metals of this group are characterized by their high reactivity with respect to the surrounding medium and by their ability to form surface compounds. Surface oxidation markedly retards the continued reduction of such metallic ions and promotes the reduction of hydrogen ions, the result being that these metals deposit out on the cathode in the form of thin oxide or hydroxide films. It is for this reason that these substances cannot be obtained in the metallic state by electrolysis.

It has already been brought out that the metallic deposit does not grow continuously over the entire surface of the cathode as electrolysis proceeds,

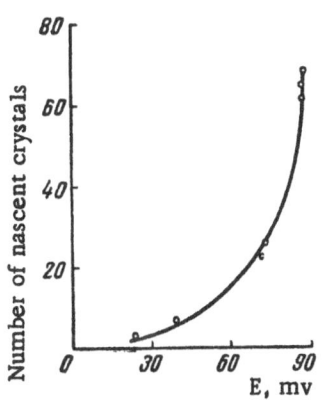

Fig. 2. The dependence of the number of nascent crystals on the electrode potential (N. T. Vagramyan).

Face-centered cubic	Body-centered cubic	Hexagonal close packed
Cu	Li	Be
Ag	Na	Mg
Au	K	Zn
Al	V	Cd
Th	Ta	Tl
Pb	α-Cr	Ti
γ-Fe	Mo	Zr
α-Co	W	Hf
Ni	α-Fe	Cr
Rh	δ-Fe	Co
Pd		Ru
Ir		Os
Pt		

but periodically, layer by layer. The adsorption of foreign particles takes place on those centers where metallic deposition has been slowed down, retarding further reduction of the metal ions and thereby altering the structure of the deposit. The amount and nature of the adsorbed particles is dependent on the composition of the electrolyte

---

* The term "chemical polarization" designates the alteration of the electrode potential which results from retardation of the reduction of metal ions during passage of the current; it can be due to dehydration, to the act of discharge itself, to the formation of two and three dimensional nuclei, and to still other factors which are not connected with concentration changes.

3

and the conditions under which electrolysis is being carried out and these factors, in turn, determine both the structure and the physicomechanical properties of the deposit. Alteration of the conditions of electrolysis affects not only the structure, but also the crystallographic form of the metallic deposit. Thus an electrolytic deposit of chromium obtained at low current density will have a lattice of the body-centered cubic type, while the deposit obtained at high current densities will show a hexagonal close packing of the atoms in a metastable form [8].

It is clear that the microstructure of the surface and the density of atomic packing are fixed by the conditions of electrolysis and that these factors, in turn, determine the physicomechanical properties of the metal.

LITERATURE

1. T. Erdey-Gruz and M. Volmer, Z. phys. Chem. 157, 165 (1931); J. Hoekstras, Rec. Trav. Chim. Pays-Bas, 50, 339 (1931), M. Volmer, Zhur. Fiz. Khim. 5, 319 (1934); A. T. Vagramyan, Zhur. Fiz. Khim. 10, 443 (1937); T. Erdey-Gruz, Z. phys. Chem. 172, 157 (1935).

2. J. N. Stranski, Z. phys. Chem. 11, 342 (1931); W. Kossel and H. Falkenhagen, Quantentheorie und Chemie (Berlin, 1928).

3. R. Kaischew, B. Mutaftschiew, and D. Nenow, Z. phys. Chem. 205, 341 (1956); R. Kaischew, E. Budewski, and J. Malinowski, Z. phys. Chem. 204, 348 (1955).

4. J. Hoekstras, Coll. Trav. chim. Tchecosl. 6, 160 (1934).

5. V. Kolschutter and A. Torricelli, Z. Elektrochem. 38, 213 (1932).

6. K. M. Gorbunova, N. A. Shishakov, and T. V. Ivanovskaya, Zhur. Fiz. Khim. 25, 981 (1951); J. B. O'Sullivan, Trans. Faraday Soc. 26, 28 (1930); J. Billiter, Principles of Electroplating [Russian translation] (ONTI, 1937); J. B. Kushner, Metall Finish. 56, 46 (1958); J. Macnaughton, J. Iron and Steel Inst. 119, 179 (1924); C. Marie and A. Buffat, J. Chem. Phys. 24, 470 (1927).

7. W. Clark, M. Burkhead, and F. Seegmiller, J. Res. Nat. Bur. Standards 39 (1947).

8. K. Sasaki and S. Sekito, Trans. Amer. Electrochem. Soc. 59, 437 (1931).

Chapter II

THE HYDROGENATION OF METALS

Crystal lattice defects, i.e., irregularities in the arrangement of atoms in the elementary cell, are quite frequently observed in electrolytic metallic deposits, their appearance being related to the specific conditions under which deposition has taken place. It has been pointed out already that this situation is due, in part, to the excitation which the ion experiences on passing through the electrical double layer prior to entering the crystal lattice. On the other hand, electrolytic growth of the metal takes place layer by layer so that at some period of the electrolysis each newly formed layer is a surface possessing higher energy than a layer in the body of the metal. Thus the force field of the surface atom is unsaturated to a certain degree and numerous foreign particles of various kinds are attracted by such atoms and included in the deposit.

Adsorption is a prime factor in fixing the structure of the metal which is formed in electrodeposition, each layer taking up hydrogen and foreign particles to an extent determined by the experimental conditions and in an amount which frequently exceeds the equilibrium solubility.

The occlusion of hydrogen in metals during electrodeposition is principally by adsorption rather than by volume absorption and differs in this respect from the occlusion which occurs when the metal is prepared by other means. It is obvious that this will lead to differences in both the amount and the rate of occlusion of the hydrogen which is taken up by the deposit.

THE ADSORPTION OF HYDROGEN AND OTHER GASES ON METALS

There is an extensive literature devoted to the adsorption of gases on the surface of a solid body [1]. Interest in the adsorption of gases dates back to the work of Berzelius in 1836 [2] and has been closely associated with the use of metals as catalysts. The study of gas adsorption has now become basic for the solution of practical problems from the most diverse fields of technology, problems such as the selection of catalysts, the protection of metals from corrosion, passivation, electroplating, etc.

The simplest case of adsorption is that which is treated in the work of Langmuir [3]. Let the fraction of the surface covered by molecules be designated by θ. The rate of evaporation of these molecules from the surface will then be proportional to the covered area, $\nu\theta$, while the rate of gas condensation will be given by

$$(1 - \Theta)\mu a,$$

$(1-\theta)$ representing the fraction of the surface which is bare, μ being the number of gas molecules impinging on unit surface in 1 sec, and α the fraction of these impinging molecules which condense.

At equilibrium $\alpha(1-\theta) \cdot \mu = \nu\theta$, or

$$\Theta = \frac{\alpha \cdot \mu}{\nu + \alpha \cdot \mu}. \tag{1}$$

The pressure is proportional to the number of molecular impacts on unit surface and it is therefore possible to write:

$$\mu = hp, \tag{2}$$

h being a coefficient of proportionality.

Substituting (2) into (1) and setting

$$\frac{\alpha \cdot h}{\nu} = k, \tag{3}$$

5

gives:

$$\Theta = \frac{\alpha h p}{\nu + \alpha \cdot h \cdot p} = \frac{kp}{1 + kp}, \tag{4}$$

where α is an exponential function of the temperature given by

$$\alpha = c e^{-\frac{Q}{RT}} \tag{5}$$

and Q is the heat of adsorption

It follows from Eq. (5) that the so-called physical adsorption should diminish with rising temperature. Experiment shows, however, that the temperature dependence of the adsorption of gases on metals is quite complex so that this simple equation is not always valid. On the contrary, it is sometimes observed that the adsorption rises with increasing temperature. From extensive experimental work, it can be concluded that there are two types of adsorption, physical and chemical.

Further study has shown the existence of various types of adsorption which are intermediate to these two extreme forms. Dubinin has expressed the belief [4] that surface compounds are formed on the adsorbent in many instances. Such compounds have characteristic chemical properties but they do not appear as free individual substances nor do they function as new phases. Frumkin [5] considers that there is an increase in the strength of bonding between the adsorbed particle and the solid surface with the passage of time. The fact that various complex relations have been observed to apply to the adsorption of gases on solid bodies can be due, at least in part, to energetic nonuniformity in the solid surface where the strength of bonding between adsorbent and adsorbed particle varies from region to region. These relations are complicated further by gas diffusion into the crystal lattice. The data on metallic hydrogen adsorption which have appeared in the literature are not entirely consistent and are frequently contradictory.

The adsorption of hydrogen on various metals has been studied recently by Kavtaradze [6] whose data shows the existence of two types of adsorption, the reversible and the irreversible. Kavtaradze has also proven that the hydrogen is instantly adsorbed on copper, silver, nickel, cadmium, and indium to form unstable surface compounds which occupy only a small fraction (4-9%) of the surface area.

His study of the temperature dependence of the adsorption of hydrogen on iron, nickel, cobalt, and platinum disclosed the existence of an adsorption hysteresis and led to the assumption that

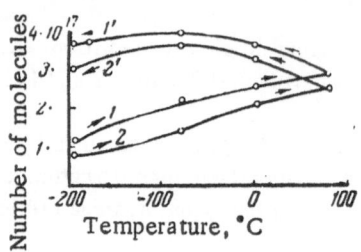

Fig. 3. The sorption of hydrogen on nickel: 1, 1') Sorption of hydrogen on a nickel film (1647 A) heated to 150° in vacuum; 2, 2') sorption of hydrogen on this same film after heating to 150° in hydrogen at $2 \cdot 10^{-2}$ mm Hg pressure with subsequent evacuation to $2 \cdot 10^{-7}$ mm Hg (Kavtaradze).

there is a reversible component of adsorption which is strongly dependent on both temperature and pressure and an irreversible component which is largely dependent on the temperature alone (Fig. 3). Some of the Kavtaradze data on the reversible and the irreversible adsorption of hydrogen on various metals at various temperatures are presented in Table 1.

This table shows that the ratio between the reversible and the irreversible components of adsorption alters radically with the temperature.

The less firmly bound hydrogen makes up about 3% of the total and can be pumped off. The remaining hydrogen represents the irreversible component of adsorption (Fig. 4a) and is firmly bound to the metal. A comparison with data obtained at higher temperatures (Fig. 4b-c) shows that the reversible component of adsorption increases markedly with rising temperature while the irreversible component diminishes.

This same author has also found the adsorption of hydrogen on nickel to be a very rapid process. Thus 91% of the hydrogen adsorption at −195° occurs in the first minute and 98.4%, in the first 10 min.

Comparison of the areas occupied by the hydrogen and nickel atoms in the surface layer at −195° showed two metal atoms per hydrogen molecule, a conclusion which is in agreement with the results of Beeck and Richter [7]. Kavtaradze considers the reversible adsorption to be due to the take-up of atomic hydrogen with the formation of compounds analogous to metallic hydrides ($Me^{+}-H^{-}$). The reversible adsorption on platinum, nickel, chromium, and iron is a molecular chemisorption of the $Me^{-}-H_2^{+}$ type.

TABLE 1. Values of the Irreversible Adsorption, $N_{irr.}$ (upper figures) and the Reversible Adsorption $N_{rev.max.}$ (lower figures) of H_2 on Iron, Nickel, and Platinum at Various Temperatures, Each Expressed as Percentage of Total Adsorption

Metal	Temperature, °C						
	−195	−78	0	50	100	150	200
Cr	92.2	93.6	84.1	70.7	—	41.3	24.7
	7.8	6.4	15.9	29.3	—	58.7	75.3
Fe	98.7	86.7	71.7	59.1	46.5	26.2	—
	1.3	13.3	28.3	40.9	53.5	73.8	—
Ni	98.4	81.7	48.9	23.0	8.6	—	—
	1.6	18.3	51.2	77.0	91.4	—	—
Pt	94.0	76.1	31.1	—	—	—	—
	6.0	23.9	65.9	—	—	—	—

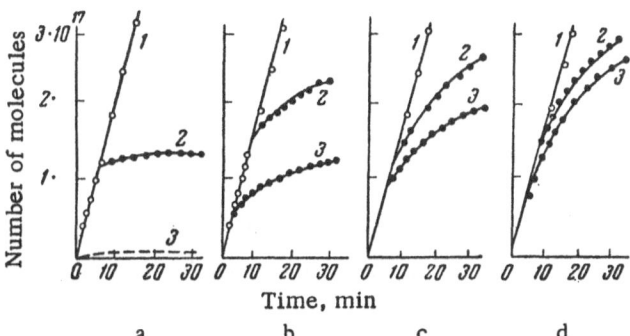

Fig. 4. The adsorption of hydrogen on a nickel film, 1662 A thick, at: a) − 95°; b) 0°; c) 50°; d) 100°. 1) Slow admission of hydrogen into evacuated reactor; 2) total adsorption; 3) reversible component of adsorption (Kavtaradze).

Various opinions have been expressed as to the form assumed by the hydrogen which is present in metals [8]. Thus Coehn and Specht [9] claim that hydrogen has protonic form in nickel, iron, and platinum and exists in the form of positively charged ions on metallic surface. These conclusions are confirmed by experiments on thermoelectric emission.

In any event, the data which have been presented show that there are at least two types of hydrogen adsorption, the one or the other predominating depending on the nature of the metal and the working conditions. Intermediate situations can also exist, and the possibility is not to be excluded that there can be various types of bonding between metal and hydrogen on a single energetically nonuniform metallic surface.

Considerable interest also attaches to the problem of the relation between the atomic weight of a metal and its hydrogen affinity. Kavtaradze has shown that the adsorption capacity of a metal is definitely related to its position in the Periodic System (Table 2). Here the strength of the hydrogen−metal bond is characterized by the heat of formation of the resulting compound. Thus the alkali and alkaline earth metals of Groups 1 and 2 form metallic hydrides while the metals of Groups 3-5 form hydrides of the intermetallic type. The solubility of hydrogen in the metals of Groups 6-10 is considerably less than the solubility in metals of the preceding groups and stable compounds are formed at both and high temperatures, the hydrogen uptake increasing as the temperature rises.

TABLE 2. Heats of Formation (Q_f), of the Hydrides of the Metals of Groups 1-5 and Heats of Atomic Chemical Adsorption of Hydrogen for Half Surface Coverage (Q_{ads}) of the Metals of Groups 6-10 in the Expanded Periodic Table

Period	Group									
	1	2	3	4	5	6	7	8	9	10
I	H Q_f									
II	Li 22					Q_{ads} ↑				
III	Na 13.9		Q_f →			← Q_{ads}				
IV	K 14.1	Ca 2.33	Sc	Ti 31.1	V	Cr 39.8	Mn (36)	Fe 33.6	Co (30)	Ni 27.0
V	Rb 13.0	Sr 21.1	Y	Zr 38.9	Nb	Mo (36-37)	Tc (32)	Ru (30-38)	Rh 23	Pd (25-26)
VI	Cs 13.5	Ba 20.5	La 40.9	Hf	Ta	W (33-34)	Re (30-31)	Os (28-29)	Ir (26-27)	Pt 25.5
VII		Ra	Ac	Th	Pa	U				

GAS DIFFUSION IN METALS

The solubility of a gas in a metal and its diffusion through the latter are both closely related to surface adsorption, and each is an important factor in electrodeposition, especially in the preparation of articles for plating. The dissolving and diffusion of gases are closely related to such effects as the swelling and hydrogen embrittlement of deposits.

The extent to which gases will penetrate into, or diffuse through, a solid body will depend on the experimental conditions. Cailletet first noted in 1863 [10] that only a part of the hydrogen evolved when an iron object is immersed in dilute sulfuric acid is actually liberated, the remainder penetrating into the metal. Similar observations are recorded in the work of Lentz (1870) [11] where it was shown that the extensive hydrogenation which accompanies electrolytic iron plating can lead to a marked change in the physicomechanical properties of a metal.

Further studies showed that not only hydrogen, but many other gases as well, are capable of penetrating various metals.

The rate of gas diffusion through a metal rises with the temperature and the pressure. The Fick linear diffusion law can be assumed to apply to the diffusion of gases through homogeneous metals:

$$\frac{dm}{d\tau} = D \cdot \frac{dc}{dx}. \tag{6}$$

Here dm is the amount of gas diffusing through unit surface in time $d\tau$, dc/dx is the concentration gradient, and D is the diffusion coefficient, a factor which depends on the nature of the metal and the diffusing gas.

The rate of change of the gas concentration is given by:

$$\frac{dc}{d\tau} = D \cdot \frac{d^2c}{dx^2}. \tag{7}$$

It is seen from Eq. (6) that the diffusion rate is proportional to the concentration gradient and therefore varies inversely with the thickness of the metallic sample under steady-state conditions.

The Effect of Temperature

According to Winkelmann [13], the rate of diffusion is related to the temperature through an equation of the form:

$$\frac{dc}{d\tau} = a \cdot T^b, \tag{8}$$

in which a and b are constants.

The experimental data on the diffusion of hydrogen through iron are well described by an equation of the type of (8) with b = 5. Lombard [14] has found the diffusion through nickel to be described by another equation, namely:

$$\frac{dc}{d\tau} = a \cdot b^T. \tag{9}$$

The temperature dependence of the diffusion rate is generally described by the equation:

$$\frac{dc}{d\tau} = d \cdot e^{-\frac{W}{T}}, \tag{10}$$

in which W is a constant and d is the slab thickness.

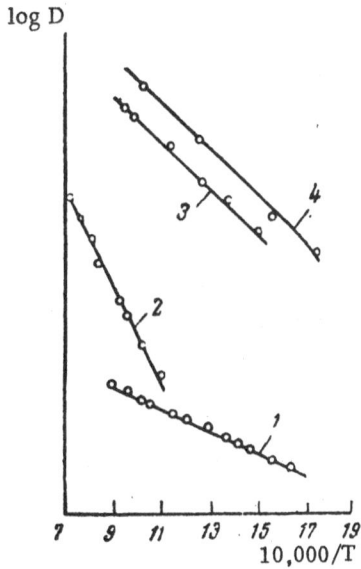

Fig. 5. The temperature variation of the diffusion rate in a number of gas—metal systems: 1) H_2—Pd, 760 mm (Lombard); 2) H_2—Pt, 274 mm (Richardson); 3) H_2—Fe, 140 mm (Smithells, Ransley); 4) H_2—Fe, 760 mm (Borelius, Lindbloom).

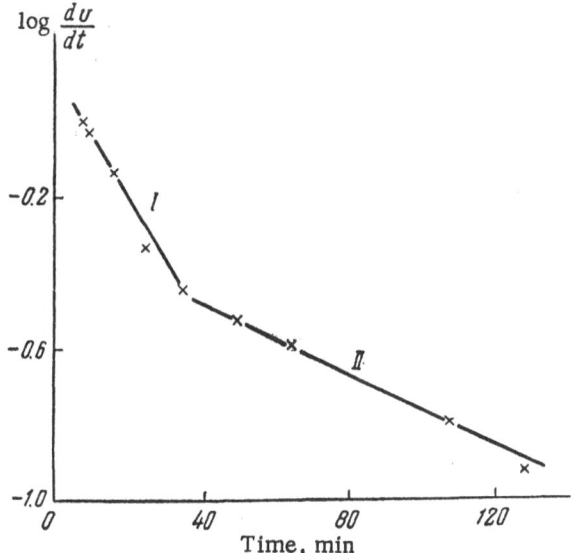

Fig. 6. Time variation of the rate of diffusion of hydrogen into iron (Smyalovskii, et al.): D_I 4.17 · 10^{-7} cm^2/sec; D_{II} 9.6 · 10^{-8} cm^2/sec.

The experimental data on the diffusion of gases through metals are in good agreement with this latter equation, representative points in a log $dc/d\tau$, 1/T plot falling nicely on straight lines (Fig. 5). The physical basis of Eq. (10) is found in the fact that the diffusion rate is inversely proportional to the activation energy, E_0. Thus one can write:

$$\log_{10} \frac{dc}{d\tau} = -\frac{E_0}{2.3\,RT} + \text{const.} \tag{11}$$

Studies extending over wide temperature intervals have shown the rate of penetration of hydrogen into metals to be temperature dependent. Thus Ham has established [15] that there is a sharp break in the plot showing the rate of hydrogen diffusion into iron as a function of the reciprocal absolute temperature over the interval from 200 to 300°, the rate of diffusion being 100 times greater above the break point than below.

Smyalovskii and his co-workers have used an original method [16] to measure the rate of diffusion of hydrogen into iron at 20° and have shown that the rate-time graph splits into two linear segments (Fig. 6). The diffusion coefficient can be calculated from this plot and proves to be equal to $4.17 \cdot 10^{-7}$ cm/sec in the initial stages, later diminishing to $9.6 \cdot 10^{-8}$ cm/sec.

These facts have served as a basis for a claim that the occlusion of hydrogen takes place in two stages, that is, that two forms of hydrogen are present in the metal.

The Effect of Pressure

Studies have shown that the rate of diffusion increases with the gas pressure, the increase being nonlinear and proportional to the square-root of the pressure (Fig. 7):

$$\frac{dm}{d\tau} = k_2 \sqrt{p}. \tag{12}$$

From this there has been developed a theory to the effect that the gas first dissociates into atoms on the metal surface and then diffuses into the interior. There is little likelihood of the diffusion of large molecules into the crystal lattice. Deviations from the relation described by Eq. (12) are observed when the diffusing gas forms a chemical compound with the metal.

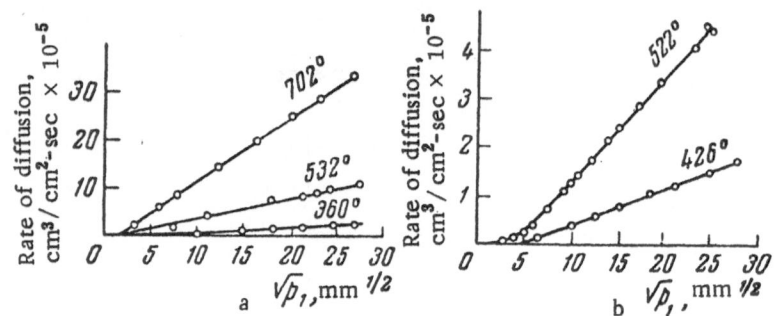

Fig. 7. Relation between the rate of hydrogen diffusion and the pressure:
a) Hydrogen—iron (Borelius, Lindbloom); b) hydrogen—nickel (Smithells, Ransley).

The Effect of the Surface State

If the major portion of the gas is assumed to diffuse through the metal in atomic form, it will follow that any variation in the surface state will alter the diffusion rate by changing the rate of breakdown of molecules into atoms. The rate of diffusion of the gas through the metal will also be affected by the degree of surface oxidation.

Smithells and Ransley [17] have shown that the rate of diffusion is independent of the pressure when the gas forms a compound on the metal surface.

Certain data showing the effect of the treatment and state of nickel and iron surfaces on the rate of hydrogen diffusion are presented in Table 3.

It is seen from this table that the diffusion rate can be altered profoundly by changing the treatment of the surface of the metal. Thus etching iron with dilute nitric acid increases the rate of diffusion almost tenfold in comparison with diffusion through a polished surface. The diffusion rate is sometimes reduced by oxidation of the metal [18].

The Effect of Alloying Substances

Geller and Tak-Ho Sun have studied [19] the effect of additions on the rate of diffusion of hydrogen through a metal, finding the diffusion coefficient and energy of activation to alter markedly with composition in the case of various alloys of α-iron and silicon. The data of these authors over the interval from 450 to 900° have been recalculated to room temperature and are presented in Table 4.

Geller and Tak-Ho Sun explain the dependence of the diffusion rate on steel composition by postulating that the presence of substances whose affinity for hydrogen is greater or less than that of iron will alter the diffusion coefficient and the activation energy for diffusion.

TABLE 3. The Effect of Surface Treatment on the Rate of Hydrogen Diffusion

Metal	Treatment	Temperature, C	Pressure, mm	Diffusion rate, cm^2/sec
Nickel	Polished	750	0.042	$1.39 \cdot 10^{-6}$
	Oxidized and reduced	750	0.042	$2.70 \cdot 10^{-6}$
	Polished	750	0.091	$2.91 \cdot 10^{-6}$
	Oxidized and reduced	750	0.091	$4.23 \cdot 10^{-7}$
Iron	Polished	400	0.770	$0.47 \cdot 10^{-7}$
	Etched (dilute HNO_3)	400	0.770	$4.40 \cdot 10^{-7}$
	Polished	520	0.073	$1.28 \cdot 10^{-7}$
	Oxidized and reduced at 600°C	520	0.073	$0.76 \cdot 10^{-7}$

TABLE 4. The Effect of Alloying Materials on the Diffusion Coeficient of Hydrogen in Steel

Material	Diffusion coefficient, cm^2/sec	Activation energy, kcal/mole H_2
Pure α-iron	$1.6 \cdot 10^{-5}$	2900
Alloy of iron with 1.06% silicon	$1.2 \cdot 10^{-6}$	4450
Alloy of iron with 1.85% silicon	$3.5 \cdot 10^{-7}$	5250
Alloy of iron with 3.11% silicon	$1.8 \cdot 10^{-8}$	7500
Alloy of iron with 4.33% silicon	$1.1 \cdot 10^{-9}$	9550

The Effect of Deposit Structure

The effect of deposit structure on diffusion is a problem of special interest and one which is closely related to the mechanism of gas diffusion through the solid body. Diffusion in crystalline bodies occurs:

1) along grain boundaries;
2) in the cleavage planes of the crystals;
3) through the crystal lattice.

Because of experimental difficulties, it has not yet been possible to obtain exact data on the effect of metallic structure on diffusion rate.

Edwards [20] has studied the rate hydrogen diffusion in mono- and polycrystalline iron deposits and has concluded that the diffusion rate is practically the same in each.

Baukloh and Kayser [21] have reached essentially the same conclusions from their study of the diffusion of hydrogen in nickel specimens of different structures. These experiments show that grain boundary diffusion is no more significant than diffusion through the entire lattice in the case of iron and nickel. On the other hand, it is known that copper heated in an atmosphere of hydrogen will become brittle as the result of partial reduction of the oxide [22]. Microstructure studies have shown that the grain boundaries are widened in such cases and it is here that most of the diffusion occurs.

Tensile tests on brittle hydrogenated metals led Pfeil [23] to conclude from the jagged rupture of the crystals that hydrogen occlusion takes place preferentially along slip planes and grain boundaries, although it is still true that a considerable portion of the gas diffuses through the crystal lattice itself.

Three mechanisms have been proposed to account for the movement of gaseous atoms through the crystal lattice [24]:

1) the atoms of gas and metal exchange places (Fig. 8a);
2) the gas atoms diffuse through the interstices between the lattice points (Fig. 8b);
3) the gas atoms diffuse along the lattice vacancies (Fig. 8c)

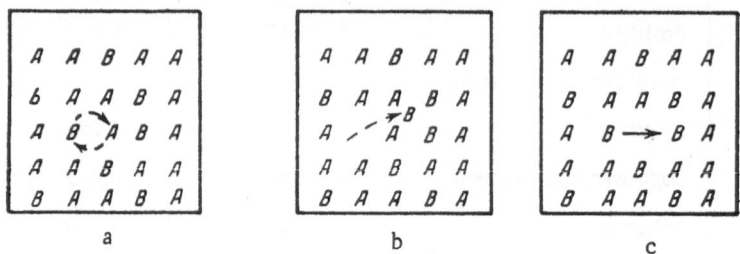

Fig. 8. Three mechanisms for gas diffusion through the crystal lattice.

The kinetic theory of gases gives certain information concerning the possibility of the movement of hydrogen through a metal by one or the other of these mechanisms.

Real Crystals and the Mechanism of Gas Diffusion through Metals

Thermal agitation and migration of the metal atoms lead to a breakdown in the regularity of distribution of atoms on the lattice of a real crystal [25]. Vaporization at temperatures less than the melting point gives indication of the movement of the atoms in the solid. It is clear that certain atoms on the surface, or even in the interior of the crystal lattice, can acquire sufficient kinetic energy to break loose completely from their neighbors. The Maxwell law shows the fraction of atoms having kinetic energy $\frac{1}{2}mv^2$ to be proportional to $e^{-\frac{1}{2}mv^2/kT}$. Thus the solid body should always contain a certain number of atoms with kinetic energies high enough to overcome the force of bonding to the neighboring atoms. The number of atoms which can acquire high energy and undergo evaporation is relatively small in solid bodies. In leaving the lattice these atoms form free lattice points (voids, holes, or dislocations). The fact that these changes are taking place, simultaneously, in various parts of the crystal makes it apparent that the real crystal will differ from the ideal in being "broken down" to a certain degree. These perturbations arise from thermal agitation and the number of dislocated atoms should, therefore, vary with the temperature, increasing as the temperature rises and reaching a maximum near the melting point.

It must be noted that two metals of different melting point will have different numbers of dislocated atoms at any one temperature, this number being lower in the metal of higher melting point. The number of vacancies can be evaluated from the energy of sublimation, U. Let the total number of atoms in a solid body be designated by N and the number of dislocated atoms by N'. It follows from the Boltzmann equation that

$$N' = N \cdot e^{-\frac{U}{kT}}.$$

Estimates on cadmium (U = 13,000 cal) show a value of N' per 1 cm^3 of 10^{-4} at room temperature and 10^{13} at 300°.

The rate of hole migration is inversely proportional to the mean time, Φ, during which the atom is located on lattice points and this, in turn, is directly proportional to

$$e^{-\frac{\Delta U}{kT}},$$

so that

$$\frac{1}{\Phi} = k_3 \cdot e^{-\frac{\Delta U}{kT}} \tag{13}$$

or

$$\Phi = A \cdot e^{-\frac{\Delta U}{kT}}, \tag{14}$$

where A is a constant.

It can be shown readily [25] that A has the value 10^{-13} sec and is therefore a quantity of the order of the period of a characteristic atomic vibration. Equation (14) can be used to calculate how long the hole will remain in a given position. The value of Φ for cadmium is 10 sec at room temperature and 10^{-4} sec at 300°, elevation of the temperature resulting in a sharp reduction of the time spent by the hole at any one point.

Frenkel' [25] claims that the dislocated atoms move to the surface of the metal through the interstices of the crystal lattice while the holes are moving to the surface along lattice points. It follows from this theory that the atoms can undergo displacement along either voids or lattice points. The probability of simultaneous exchange of two atoms, one moving into the place of the other, is quite low, however. The kinetic theory of the solid body brings out clearly the effect of temperature and the nature of the metal on the mechanism of atomic diffusion through the crystal lattice.

Thus the work on the diffusion of hydrogen through metals gives reason for supposing that the diffusion rate depends on both the nature and the structure of the metal. The diffusion of hydrogen follows various paths (along grain boundaries and lattice and crystal defects) and the metal can therefore be assumed to contain various forms of hydrogen. This conclusion is confirmed by differences in the measured values of the diffusion coefficient. The existence of various types of hydrogen—metal bonding can be brought out by studying the rate of evolution of gas from the vacuum heated deposit. Thus Popova and Gorbunova [26] have shown that a large portion of the hydrogen in a manganese deposit is liberated by heating to 125° and that practically all of it can be removed by heating to 300° (Fig. 9, Curve 1).

The rate of liberation of hydrogen is considerably lower for nickel than for manganese, appreciable evolution beginning only above 300° in this case (Fig. 9, curve 2).

From these studies, Popova and Gorbunova concluded that hydrogen exists in nickel in the form of a penetrating solid solution whose composition alters irregularly on moving into the interior of the deposit while the hydrogen in manganese is distributed along grain and block boundaries.*

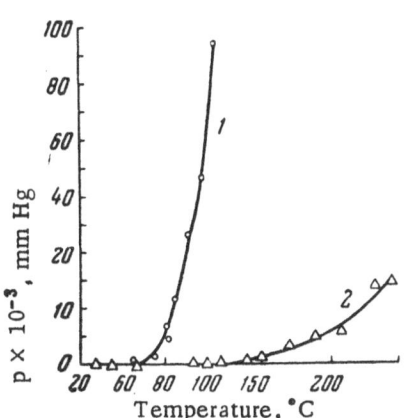

Fig. 9. The relation between the temperature and the amount of hydrogen evolved from an electrolytic deposit: 1) Manganese obtained from an electrolyte containing $MnSO_4 \cdot 5H_2O - 150\text{-}200$ g/liter + $(NH_4)_2SO_4 - 50\text{-}100$ g/liter; 2) nickel obtained from an electrolyte containing $NiSO_4 \cdot 7H_2O - 130$ g/liter + $H_3BO_3 - 30$ g/liter + $NaCl - 3$ g/liter + $NaF - 5$ g/liter $- 50\text{-}100$ g/liter.

THE SOLUBILITY OF HYDROGEN IN METALS

When applied to gases and metals, the term "solubility" usually designates the amount of gas taken up at equilibrium. Here, account is to be taken of the existence of such various types of combination between metal and gas as solid solutions and chemical compounds. Each conceivable type of gas—metal equilibrium will be established here and it is, therefore, obvious that the hydrogen distribution will vary with the structure and degree of purity of the metal, equilibrium in the intercrystalline planes differing from equilibrium in the lattice itself and from equilibrium in regions containing foreign inclusions.

The uptake of hydrogen in cases of extensive compound formation between hydrogen and the metal is much greater than would correspond to simple dissolution.

The experimental determination of the solubility of gases in metals is often a difficult problem and the data which have been obtained are frequently contradictory.

Major interest attaches to a study of the effect of temperature and pressure on solubility.

Sieverts and Borelius [27] claim that the dependence of solubility on temperature is expressed through an equation of the form:

$$B_P = c \cdot e^{-\frac{E_S}{2RT}}, \qquad (15)$$

in which c is a constant and E_S is the heat of solution.

---

* The higher rate of evolution of hydrogen in the case of manganese can also be related to the existence of a well-defined network of cracks.

It follows that this equation should graph as a straight line in the coordinates $\log B_p$, $1/T$ and the experimental results on a number of metals actually fulfill this condition rather well (Fig. 10).

The heat of solution can be determined from the slope of the $\log B_p$ vs. $1/T$ line. Values of E_S (cal/mole H_2) for various metals are given below.

Aluminum	45,500
Silver	11,600
Copper	28,600
Nickel	5,400
Cobalt	17,000
Iron	15,000

It must be pointed out that this temperature-solubility relation is not valid for all metals, a fact which is seen most clearly in the results from a study of the solubility of hydrogen in titanium, zirconium, thorium, and vanadium (Fig. 11). The figure makes it clear that the solubility here diminishes with rising temperature. Such anomaly is explained in principle by the fact that dissolved hydrogen forms hydrides with these metals. It must be noted, however, that one is not always justified in drawing such clear distinctions regarding the metal–hydrogen bond since there are cases in which an alteration in the experimental conditions will bring about a passage from one type of bonding to the other.

Thus the work of Balandin, Pecherskaya, and Stakhov [28] has shown that nickel forms a series of hydrides (formulas: NiH, NiH_2, and NiH_4) which can be isolated and studied.

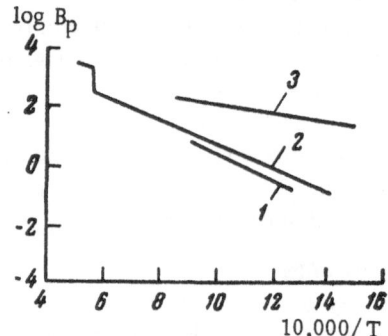

Fig. 10. The effect of temperature on the solubility of hydrogen in metals: 1) Cobalt; 2) iron; 3) nickel (Smithells).

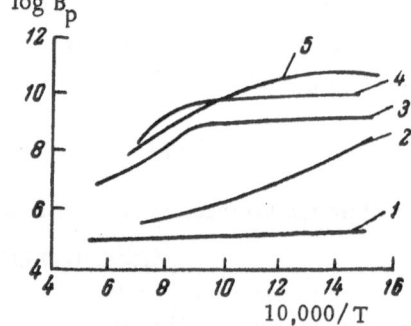

Fig. 11. Adsorption isobars for hydride forming metals: 1) Palladium; 2) vanadium; 3) thorium; 4) zirconium; 5) titanium (Smithells, Fowler).

Studies on the effect of temperature have brought out the interesting fact that the amount of dissolved hydrogen increases sharply at the melting point of the metal or at a transition point (Fig. 12). Thus the amount of hydrogen occluded in copper rises sharply at the melting point. It has also been observed that the solubility of hydrogen is greater in γ-iron than in α-iron, and similar effects are noted with the various modifications of chromium.

Studies on the effect of foreign materials on the solubility of gases have proven that the amount of gas occluded in copper can be altered profoundly by the presence of alloying metals (Fig. 13), the figure showing that this solubility is increased markedly by the addition of 20% nickel and reduced by the addition of aluminum.

It has been pointed out above that a rise in the gas pressure can markedly increase the diffusion into metals and thus lead to a rise in the solubility. Here the solubility increases in proportion to $\sqrt{p}$ (Fig. 14). It is to be seen from the figure that iron, nickel, cobalt, and certain other metals deviate from this rule at low pressures. The rule is, furthermore, not in general applicable to zirconium, titanium, and vanadium, a fact which is to be explained by the formation of chemical compounds between these metals and hydrogen.

In principle, all metals can be classified into three groups on the basis of their affinity for hydrogen.

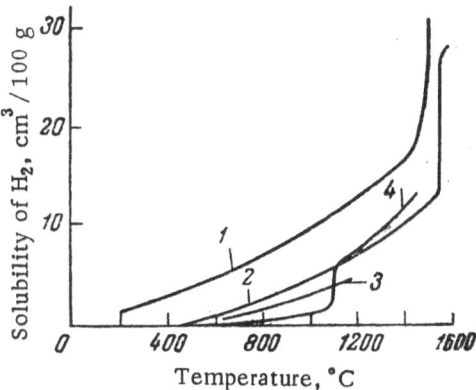

Fig. 12. The solubility of hydrogen in nickel (1), iron (2), cobalt (3), and copper (4) (Smithells).

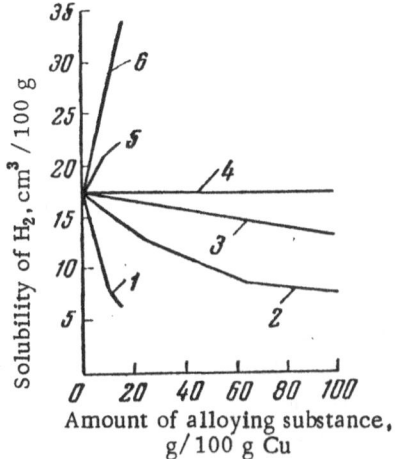

Fig. 13. The affect of the addition of aluminum (1), tin (2), gold (3), silver (4), platinum (5), and nickel (6) on the solubility of hydrogen in copper (Sieverts).

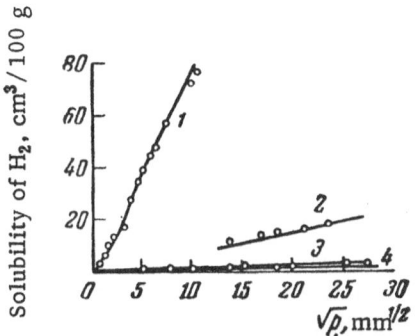

Fig. 14. Isotherms for the adsorption of hydrogen on: nickel at 600° (1), iron at 1000° (2), cobalt at 1000° (3) and silver at 300° (4) ($\sqrt{p}$ relation, Smithells).

TABLE 5. The Effect of the Nature of the Metal on the Hydrogen Solubility

Temp. °C	Hydrogen solubility, cm^3/g metal									
	Nb	Pd	Th	Ta	Ti	V	Zr	Ni	Fe	Co
20	55	—	125	—	403	150	—	≪ 0.05	≪ 0.05	≪ 0.05
300	44	3.3	107	34	384	60	270	—	—	—

The first of these groups contains those metals in which the hydrogen solubility is infinitesimally low, or zero, at room temperature; namely: lead, cadmium, tin, zinc, silver, copper, and certain others.

The second group contains iron, nickel, cobalt, chromium, manganese, and other metals in which the solubility of hydrogen is low, but measurable, at ordinary conditions.

The third group contains those metals which have the ability to form compounds with hydrogen and dissolve considerable quantities of the gas even at room temperature. In this group are found niobium, titanium, zirconium, vanadium, thorium, and certain other metals.

Table 5 gives the solubility of hydrogen in certain metals from these various groups.

It will be shown, in what follows, that the hydrogen content of an electrolytic metal does not correspond to equilibrium, but is considerably higher than the above figures would indicate.

LITERATURE

1. Collection, Surface Chemical Compounds and Their Role in Adsorption Phenomena [in Russian] (Moscow State University Press, 1957); B. M. Trapnell, Chemisorption [Russian translation] (Foreign Literature Press, Moscow, 1958); S. Z. Roginskii, Adsorption and Catalysis on Nonuniform Surfaces [in Russian] (Acad. Sci. USSR Press, Moscow, 1948).
2. Berzelius, Jahr. f. Chem. 15, 237 (1836).
3. I. Langmuir, Schweiggers. J. 34, 91 (1922).
4. M. M. Dubinin, Collection, Surface Chemical Compounds and Their Role in Adsorption Phenomena [in Russian] (Moscow State University Press, 1957) p. 9.
5. A. N. Frumkin, Surface Chemical Compounds and Their Role in Adsorption Phenomena [in Russian] (Moscow State University Press, 1957) p. 53.
6. N. N. Kavtaradze, Zhur. Fiz. Khim. 32, 909 (1958); Doklady Akad. Nauk SSSR, 114, 822 (1957); Heterogeneous Catalysis [in Russian] (Goskhimizdat, Moscow, 1955) p. 150; Abstracts of Reports, Conference on Catalytic Hydrogenation and Oxidation (Alma-Ata, 1954) p. 24; Surface Chemical Compounds and Their Role in Surface Phenomena [in Russian] (Moscow State University Press, 1957) p. 73; N. N. Kavtaradze, Zhur. Fiz. Khim. 32, 1055, 1214 (1958); N. N. Kavtaradze, Izvest. Akad. Nauk SSSR, Otdel Khim. Nauk 9, 1045 (1958).
7. O. Beeck and A. W. Richtel, Disc. Faraday Soc. 8, 159 (1950).
8. J. G. de Boer, Electron Emission and Adsorption Phenomena [Russian translation] (ONTI, Moscow-Leningrad, 1936).
9. A. Coehn and W. Specht, Z. Phys. 62, 1 (1930); A. Coehn and H. Jurgens, Z. Phys. 71, 179 (1931).
10. L. Cailletet, Compt. rend. 56, 847 (1863).
11. R. Lentz, Zhur. Fiz. Khim. Obshchestvo 2, 57 (1870); Bull. de l'Academ. Imperiale des Sciences de St.-Petersbourg 14, 337 (1870).
12. F. Seitz, The Physics of Metals [Russian translation] (Gostekhizdat, Moscow-Leningrad, 1947); C. S. Smithells, Gases and Metals [Russian translation] (Metallurgizdat, Moscow, 1940); R. Barrer, Diffusion in and through Solid Bodies [Russian translation] (Foreign Literature Press, Moscow, 1948).
13. A. Wikelmann, Ann. phys. 6, 104 (1901); 17, 591 (1905); 19, 1045 (1906).
14. V. Lombard, Compt. rend. 177, 116 (1923).
15. W. Ham, J. Chem. Phys. 1, 476 (1933).
16. M. Smyalovskii, B. Baranooskii, Z. Shklyarskaya-Smyalovskaya, K. Skalch'mooskii, G. Angershtein, A. Kunstetter, M. Zhubr and G. Yarmolovich, Soviet Electrochemistry (Consultants Bureau, New York, 1961) Vol. 1, p. 109.
17. C. Smithells and C. Ransley, Proc. Roy. Soc. A, 152, 706 (1935).
18. C. Smithells and C. Ransley, Proc. Roy. Soc. A, 157, 292 (1936).
19. W. Geller and Tak-Ho Sun, Arch. Eisenhüttenwesen 281, 423 (1950).
20. G. Edwards, J. Iron and Steel Inst. 110, 9 (1924).
21. B. Baukleh and H. Kayser, Z. Metallkunde 27, 281 (1935).
22. F. Rhines and C. Mathewson, Trans. Inst. Min. (Metall) Engrs. 111, 337 (1934).
23. J. B. Pfeil, Proc. Roy. Soc. 112, 182 (1926).
24. F. Seitz, The Physics of Metals [Russian translation] (Gostekhizdat, Moscow, 1947).
25. Ya. I. Frenkel', Introduction to the Theory of Metals [in Russian] (Gostekhizdat, Moscow, 1948).
26. O. S. Popova and K. M. Gorbunova, Zhur. Fiz. Khim. 32, 2020 (1958).
27. A. Sieverts, Z. Metallkunde 21, 37 (1929); Borelius, Ann. phys. 16, 174, 190 (1933).
28. A. A. Balandin, B. V. Pecherskaya, and K. A. Stakhov, Zhur. Organ. Khim. 11, 557 (1941).

Chapter III

METHODS OF STUDYING THE HYDROGENATION
OF ELECTROLYTIC DEPOSITS

THE METHOD OF VACUUM HEATING

The method of vacuum heating is the oldest of the procedures used for determining the amount of hydrogen occluded in an electrolytic deposit and the one which is generally employed [1]. In essence, this involves the introduction of the deposit into a closed system where it is evacuated to a pressure of the order of 10^{-6} mm Hg and held at fixed temperature for as long as may be necessary. The elimination of gases and vapors from the deposit brings about an increase in the pressure in the system and the volume of the evolved gases can be determined from the difference in the pressures before and after heating. It is to be noted that this procedure determines not only hydrogen but water vapor and the decomposition products which arise from thermal breakdown of the organic compounds and foreign substances which are present in the deposit.

An apparatus for the determination of the amount of hydrogen taken up by a deposit is represented schematically in Fig. 15. This system consists of the diffusion pump A, a fore-vacuum oil pump which is not shown in the figure, the MacLeod gauge B for pressure measurements, and the quartz glass cell C which is set in the tubular furnace 3. This cell is usually of the type shown at 1 and is constructed from quartz glass because of the high thermal stability of this material; into it, the working specimen is loaded. The cell is surrounded by the sleeve 2 which is connected to the vacuum line by a ground-glass joint, the aim being to maintain high vacuum in the system despite the permeability of quartz to gases at high temperature. The bent tube 4 is attached to one end of the cell by a ground joint, while a second joint connects the cell to the vacuum line through the U-tube 5. This U-tube is designed to freeze out moisture and the easily condensable decomposition products which result from heating the substances occluded in the deposit during electrolysis. The hydrogen can be separated out of the gaseous mixture and determined with a certain degree of accuracy by using liquid nitrogen for cooling. The specimen is heated by the tubular furnace 3 which has been previously calibrated.

The determination of the amount of gas occluded in a deposit is made in the following manner. It is first proven that the system is working normally and that air does not leak into it when the vacuum pumps are cut off. The cell 1 is then cut off from the rest of the line by the cocks 7 and 8, the tube 4 is removed, carefully, and the specimen placed in the bottom of it. This tube is then reattached and the lower end containing the sample is inserted in the Dewar vessel 6. This vessel is filled with liquid air or nitrogen, and serves to prevent premature loss of gases and water vapor during preliminary evacuation. The system is then reevacuated, cock 7 being opened with great care to avoid breaking the MacLeod gauge by the rush of the air from the cell. Pumping is stopped when the desired vacuum has been attained. The Dewar flask is then removed from tube 4 and the latter rotated so as to transfer the specimen into the quartz tube 1; at the same time, the U-tube 5 is immersed in the Dewar flask. Finally, the electrical furnace is turned on to heat the specimen.

It should be noted that only cock 8 (which connects the cell sleeve with the forepump) is open during heating. The system is cooled to room temperature at the end of the heating period and the pressure measured on the MacLeod gauge. From the difference between the pressure prior to, and after heating, the volume, v, of evolved gases can be calculated from the equation:

$$\frac{pv}{T} = \frac{p_0 v_0}{T_0},$$

in which p is the pressure after the specimen has been heated, T is the absolute temperature, v_0 is the volume of the system (previously determined), p_0 is the normal pressure (760 mm Hg), and T_0 is the normal temperature (273°).

The water and easily condensible components can be determined by removing the flask with the liquid air or nitrogen from the U-tube and measuring the pressure once more after the condensed mixture has evaporated. The amount of vapors and gases is calculated from the pressure difference.

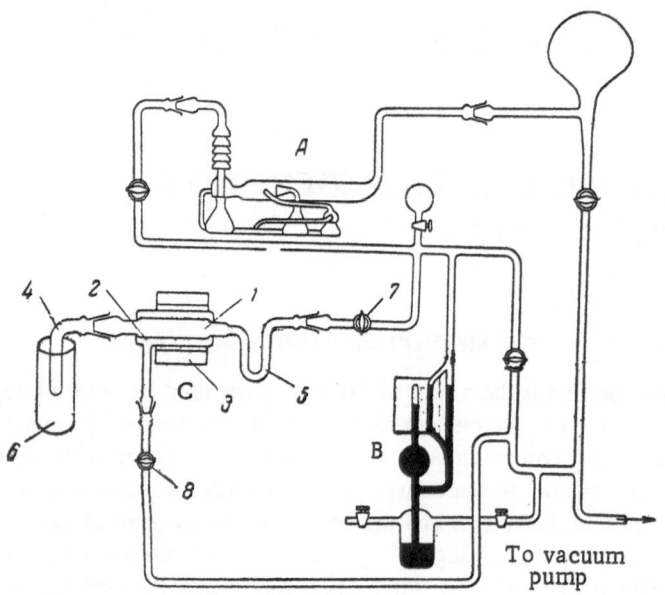

Fig. 15. Schematic diagram of a vacuum system for measuring the hydrogenation of electrolytic deposits.

The time required for complete removal of the gases from the specimen is determined at each working temperature by developing a relation between the time of heating and the amount of evolved gases.

The amount of gas evolved by vacuum heating gradually increases with the passage of time and eventually reaches a constant value. The time required for reaching this constant value varies with the temperature and diminishes as the temperature rises. It has been observed, repeatedly, that this limiting volume corresponds to the complete elimination of gases from the specimen. Thus it is necessary to establish the time and temperature required for the complete degasification of each metal. For example, heating for 1.5-2 hr at 500-600°C is adequate for the complete removal of gases from electrolytic deposits of nickel, chromium, iron, and zinc.

THE RHEOMETER METHOD

The essential feature of this method is that the gases evolved from the vacuum-heated specimen are passed through a capillary tube whose parameters are known [2]. The rate of gas flow can be obtained by measuring the pressure at the two ends of the capillary tube and making use of the equation:

$$v = k \cdot \frac{\Delta p}{f_t},$$

in which v is the volume of gas passing through the capillary in unit time, k is a constant for the apparatus, Δp is the difference of the pressures at the ends of the capillary, and f_t is the coefficient of internal friction of the moving gas.

The specimen is heated to a fixed temperature in a quartz tube which is connected to a vacuum line. One end of this quartz tube is equipped with a thermocouple, while the other end is connected to a rheometer which is more or less completely filled with kerosene, water, mercury or some such liquid. The rate of evolution and volume of the gases can be determined from a measurement of the pressure difference, Δp, in the rheometer. It must be noted that it is necessary to calibrate the rheometer in order to allow for alterations in the internal friction in passing from one gas to another.

It is also to be noted that the rheometer method suffers from the fact that the pressure rise in the system is due not only to the evolution of gases but also to the alteration in the volume of these gases resulting from a non-uniform elevation of the temperature. It is very difficult to allow for this effect.

THE MOLTEN METAL METHOD*

The vacuum system for the determination of gases by the melt method uses a different type of furnace and a different procedure for gas analysis.

The metal under study is placed in a graphite crucible and heated to complete fusion in a tubular induction furnace. A mercury diffusion pump carries away the gases evolved in fusion and a Toepler pump collectes these gases in a vessel of known volume where the mole number is determined from the pressure reading on a MacLeod gauge [3].

This method is applicable only to metals of low, or moderately high, melting points. Vacuum methods for determining occluded gases suffer from the common defect that the air in the system must be pumped away before the specimen is heated and an indeterminate amount of gas is thus invariably lost. Morozov [4] has developed an apparatus in which this source of error is eliminated by introducing the specimen through a mercury seal.

Turovtseva et al. [5] have proposed a variant of this apparatus which has a dead space correction of 1 ml for 15 min at 1800° and gives determinations which are accurate to $1 \cdot 10^{-4}\%$. The furnace used here is so designed that the specimen can be introduced under vacuum. This apparatus also permits hydrogen to be separated from other gases by diffusion through a palladium filter.

THE METHOD OF DEFORMATION OF THE METALLIC CATHODE

An original method for the determination of hydrogenation proposed by Smyalovskii [6] is based on the fact that the metallic cathode will be elongated by the uptake of hydrogen to an extent which is dependent on the working conditions. The measurements involve a determination of the increase in length of the cathode, the sensitivity being increased by making the cathode in the form of a seven-loop spiral. This spiral is initially 22 mm in diameter and is placed under weak tension by a small glass bob hung on its lower end. This tension is without appreciable influence on the deformation of the cathode, calculations based on the cross section of the wire showing its value to be of the order of 2.5 g/mm². The elongation of the cathode is observed through a cathetometer, cathodic polarization being continued until a stationary state is reached and the cathode is saturated with hydrogen.

LITERATURE

1. W. Newel, J. Iron and Steel Inst. 141, 243 (1940); S. S. Nasyrova and G. I. Chufarov, Zavodskaya Lab. 11-12, 1047 (1945).
2. Yu. V. Baimakov and M. I. Zamotorin, Proceedings of a Conference on Electrochemistry [in Russian] (Acad. Sci. USSR Press, Moscow, 1953) p. 125.
3. S. Dushman, High Vacuum Techniques [Russian translation] (Foreign Literature Press, Moscow, 1950).
4. A. N. Morozov, Hydrogen and Nitrogen in Steel [in Russian] (Metallurgizdat, Moscow, 1950).
5. Z. M. Turovtseva, N. F. Litvinova, G. V. Mikhailova, A. S. Noskova, and R. Sh. Khalitov, Zhur. Anal. Khim. 12, 208 (1957).
6. M. Smyalovskii and Z. Shklyarskaya-Smyalovskaya, Byull. Pol'sk. Akad. Nauk, Otdel 3, 2, 73 (1954); Izvest. Akad. Nauk SSSR, Otdel Khim. Nauk 225 (1954).

---

* The liberation of carbon monoxide and nitrogen and the necessity for additional gas analysis are complicating factors in the determination of hydrogen at temperatures in excess of 650°.

Chapter IV

THE EFFECT OF VARIOUS FACTORS ON THE
HYDROGENATION OF THE METALLIC DEPOSIT

The hydrogenation of the electrolytic deposit is an important problem and one which has yet to be studied systematically. The fact that the reported data are frequently fragmentary, or even contradictory, is probably to be traced back to the use of involved and inexact methods for measuring hydrogenation. It is equally true that the quality of the deposit is not taken into account in determining the hydrogen occlusion although it is a significant factor in fixing the hydrogen content.

One of the principal objections to the work which has been done in this field is that it has generally involved determination of the total occlusion of gas without giving consideration to the amounts of the various component substances. Furthermore, no account has been taken of the fact that the effect of a gas on the mechanical properties of a deposit will depend on whether this gas is present as molecules, adsorbed atoms, or atoms included in the crystal lattice. The absence of exact data makes for difficulty in giving a definitive treatment of the effect of hydrogenation on the physical chemical and mechanical properties of the deposit.

THE EFFECT OF THE NATURE OF THE METAL

The material found in the text of Billiter [1] shows the following relationship to exist between the degrees of hydrogenation of the electrolytic deposit and the nature of the cathode.

TABLE 6. The Hydrogenation of Various Metals

Metal	Hydrogenation, volume%*	Investigator
Tin	5-12	Smithells
Nickel	79	Gossenbrook
"	625	Krausmann, Billoni, Lantoni
Iron	180	Hughes
"	780-6000	Krausmann, Billoni, Lantoni
Zinc	2000	Schwartz
Chromium	180,000-260,000	Girard

* The values presented here have been obtained in the following manner. Assume that there is 0.70 cm^3 of occluded hydrogen in 1 g of an electrolytic nickel deposit. If 1 g of nickel is supposed to have a volume of 0.112 cm^3, it follows that:

$$\frac{0.70}{0.112} \times 100 = 625\%.$$

TABLE 7. The Effect of the Nature of the Metal on Hydrogenation

Metal	Conditions of preparation	Amt. of occluded gas, cm^3/g	Amt. of occluded hydrogen, cm^3/g	Investigator
Pd	Ammonium chloride electrolyte $I_c = 0.5-2.0$ amp/dm^2	—	0.33-0.83	Vagramyan
Cr	CrO_3-250 g/liter; $H_2SO_4-2.5$ g/liter; $I_c = 10-50$ amp/dm^2	—	1-10.0	Tsareva-Petrova
Ni	1 N $NiSO_4 \cdot 7H_2O$; 25°; pH 1.5-2.0; $I_c = 2.0$ amp/dm^2	0.81-2.2	0.1-0.58	Tsareva-Petrova
Co	1 N $CoSO_4 \cdot 7H_2O$; 25°; pH 2.0-4.0; $I_c = 2.0$ amp/dm^2	1.3-6.4	0.7-4.3	Tsareva-Petrova
Fe	1 N $FeSO_4 \cdot 7H_2O$; 25°; pH 1.8-3.2; $I_c = 2.0$ amp/dm^2	4.1-5.76	2.3-4.3	Tsareva-Petrova
Cu	$CuSO_4 \cdot 5H_2O-250$ g/liter; 20-25°; H_2SO_4-50 g/liter; $I_c = 1.0-3.0$ amp/dm^2	0.2-0.28	0.036-0.086	Tsareva-Petrova
Mn	$MnSO_4 \cdot 5H_2O-150-200$ g/liter; $(NH_4)_2SO_4-50-100$ g/liter; pH 3.8-4.0; $I_c = 10-50$ amp/dm^2	—	5-8	Popova, Gorbunova
Zn	ZnO—1 g-equiv/liter NaCN—2 g-equiv/liter NaOH—2.5 g-equiv/liter	—	0.22	Kudryavtsev, Moroz

Table 6 shows that the degree of hydrogenation diminishes in passing from chromium to tin.

It is unfortunate that no indication is given of the electrolytes which were used in preparing these deposits or the conditions of electrolysis. The hydrogen content of the deposit is, however, quite dependent on such factors, so that the above sequence would not apply under all experimental conditions.

The gas occlusion can also be closely dependent on the anion of the salt from which the metal is deposited. Thus different quantities of occluded gases will be present in deposits of metals of the iron group obtained from chloride and from sulfate solutions. Specifically, the cobalt deposit obtained from a chloride electrolyte at $I_c = 2.0$ amp/dm^2 and pH 2.5 contains 0.3 cm^3/g of hydrogen whereas the deposit obtained from a sulfate electrolyte under the same conditions contains 4.3 cm^3/g.

The occlusion of gas in a metal will depend not only on the anion of the electrolyte but also on the foreign substances which may be present. Thus an electrolytic iron deposit obtained from 1 N $FeSO_4$ at $I_c = 2$ amp/dm^2 and pH 2.5 contains 3.64 cm^3/g of occluded hydrogen. This hydrogen occlusion can be raised to 3.74 cm^3/g by adding aluminum sulfate at a concentration of 100 g/liter to the solution. The addition of 30 g/liter of boric acid markedly increases the amount of hydrogen which can be driven out of the deposit during vacuum heating, raising this figure to 12 cm^3/g. These data show that the composition of the electrolyte and the conditions of electrolysis must both be taken into account in treating the effect of the metal on hydrogenation. This is not always possible, however, since cases exist in which deposits of identical quality cannot be obtained from the same type of electrolyte.

Table 7 presents recent data of various authors on the hydrogenation of metals.

These figures show that hydrogen occlusion is greatest in chromium deposits and diminishes in the sequence Cr > Mn > Fe > Co > Ni > Zn > Cu.

The extremely low value for palladium is due to the fact that the vacuum determination of occluded hydrogen involves the loss of a large and indeterminate fraction of this gas prior to heating the specimen.

The above sequence for the alteration in hydrogenation in passing from iron to nickel in the iron group is in complete accord with the data of Foerster [2].

THE EFFECT OF CURRENT DENSITY

There are comparatively few data on the effect of current density on hydrogenation and these are, in part, contradictory. The hydrogen occlusion in nickel deposits generally increases with a rise in the current density. Figure 16 shows the relation between the amount of hydrogen occluded in a chromium deposit and the current density [3].

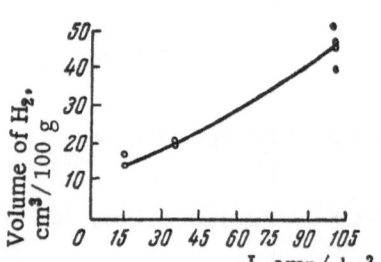

Fig. 16. The relation between the current density and the amount of hydrogen present in a chromium deposits (Rykova). Deposits obtained from an electrolyte containing 250 g/liter of CrO_3 and 2.5 g/liter of H_2SO_4 at 65°.

Here, too, there is an increase in the hydrogen occlusion with rising current density.

An increase in the current density increases the surface concentration of hydrogen atoms, thereby raising the probability of the occlusion of hydrogen in the deposit. It should be noted, however, that the degree of hydrogenation of the deposit is not invariably increased by a rise in the current density, since there are cases in which such increase will sensibly lower the quality of the deposit and thus reduce the uptake of hydrogen.

The occlusion of hydrogen in electrolytic manganese deposits passes through a maximum as the current density is increased (Table 8 [4]). It is probable that the reduction of quality of the deposit accounts for this diminution of hydrogenation with increasing current density.

THE EFFECT OF TEMPERATURE

Experiment has shown that the temperature of the electrolyte is one of the principal factors in fixing the occlusion of hydrogen by a deposit.

TABLE 8. The Effect of Current Density on the Hydrogenation of Manganese

Solution composition	Current density, amp/dm²	Amount of hydrogen, cm³/g
MnSO₄ · 5H₂O— 150-200 g/liter	20	4.8
	30	5.7
(NH₄)₂SO₄—50-100 g/liter	50	5.4
	20	7.0
The same	30	7.7
20 cm³/liter glycerine	50	6.2

Thus the data of Foerster [2] and Lentz [5] show that the amount of hydrogen present in a metal deposit from the iron group is sharply reduced by increasing the temperature of the electrolyte from 18 to 75°.

Temperature °C	18	37	55	75
Volume of hydrogen cm³/g iron	9.52	3.92	2.67	1.07

These same authors reported similar results in a study of the electrodeposition of nickel from a solution containing 260 g/liter of $NiSO_4 \cdot 7H_2O$, 30 g/liter H_3BO_3, 24 g/liter Na_2SO_4, and 20 g/liter $MgSO_4$ (Fig. 19).

The figure makes it clear that an elevation of temperature from 20 to 63° reduces the hydrogen occlusion to less than one-third of its original value.

A compilation of data on the effect of temperature on the hydrogenation of the metals of the iron group is given in Table 9.

It is seen from this table that an elevation of temperature favors a diminution in the amount of gas occluded by these metals. The fact that the overvoltage also increases is clear indication that increased hydrogenation of the deposit results in a retardation of the reduction of metal ions.

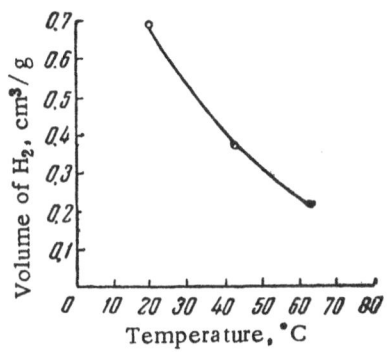

Fig. 17. The relation between the electrolyte temperature and the hydrogen occlusion in a nickel deposit (Vagramyan, Petrova). Deposit obtained from an electrolyte having the composition: $NiSO_4 \cdot 5H_2O - 260$ g/liter; $H_3BO_3 - 30$ g/liter; $Na_2SO_4 \cdot 10H_2O - 24$ g/liter; $MgSO_4 \cdot 7H_2O - 20$ g/liter; pH = 3.78; t = 18-20°; I_c = 2.0 amp/dm²

THE EFFECT OF THE pH OF THE ELECTROLYTE

The amount of gas occluded by the metals of the iron group alters radically with the pH of the electrolyte and this factor must, therefore, be taken into account in carrying out electrodeposition of these substances. Figure 18 illustrates this situation for an iron deposit obtained from 1 N $FeSO_4$ at I_c = 2.0 amp/dm² and 25°.

The effect of pH on the hydrogen occlusion is shown in Curve 1, while Curve 2 shows the effect of this same factor on the occlusion of those gases which condense at the temperature of liquid nitrogen. These latter gases are composed of water from the decomposition of metallic hydroxides, from the reduction of metallic oxides during vacuum heating, and from mechanical occlusion in the deposit, and carbon monoxide and carbon dioxide from the decomposition of traces of organic substances which enter the deposit from the electrolyte [6]. Analysis of experimental results shows that the amount of these gases generally increases with the pH of the solution and from this it can concluded that they arise, for the most part, from the occluded hydroxides. The curves show that the hydrogen occlusion diminishes sharply with a rise in the pH of the solution, falling from 4.3 cm³/g at pH 1.8 to 2.3

TABLE 9. The Relation between the Temperature and the Gas Occlusion in Deposits of the Metals of the Iron Group

Composition of electrolyte and conditions of electrolysis	Temperature, °C							
	25			50		75		
	H_2 content, cm³/g	total gas occlusion, cm³/g	over-voltage, mv	H_2 content, cm³/g	over-voltage, mv	H_2 content cm³/g	total gas occlusion, cm³/g	over-voltage, mv
Fe 1 N FeSO; pH 1.9; I_c = 2.0 amp/dm²	4.35	5.75	396	—	—	0.7	1.33	200
Ni 1 N $NiSO_4$; pH 1.9; I_c = 2.0 amp/dm²	0.95	2.58	560	0.35	445	0.35	0.95	320
Co 1 N $CoSO_4$; pH 1.9; I_c = 2.0 amp/dm²	0.86	2.36	188	—	—	0.2	0.41	211

cm³/g at pH 3.2, while the occlusion of considerable gases (and hydroxides) increases somewhat, ranging from 1.34 cm³/g to 1.5 cm³/g at these same pH values.

It is also true that the occlusion of hydrogen is much more extensive than the occlusion of hydroxides over this pH interval. It is interesting to compare the volume of gas occluded in iron deposits obtained from common electrolytes with the volume occluded in deposits obtained from electrolytes containing both iron and aluminum sulfates (Fig. 19). The figure shows the same type of variation of occluded hydrogen and hydroxides with the pH of the solution as in the preceding case, the hydrogen diminishing rapidly (Curve 1), and the hydroxides increasing (Curve 2), as the pH rises.

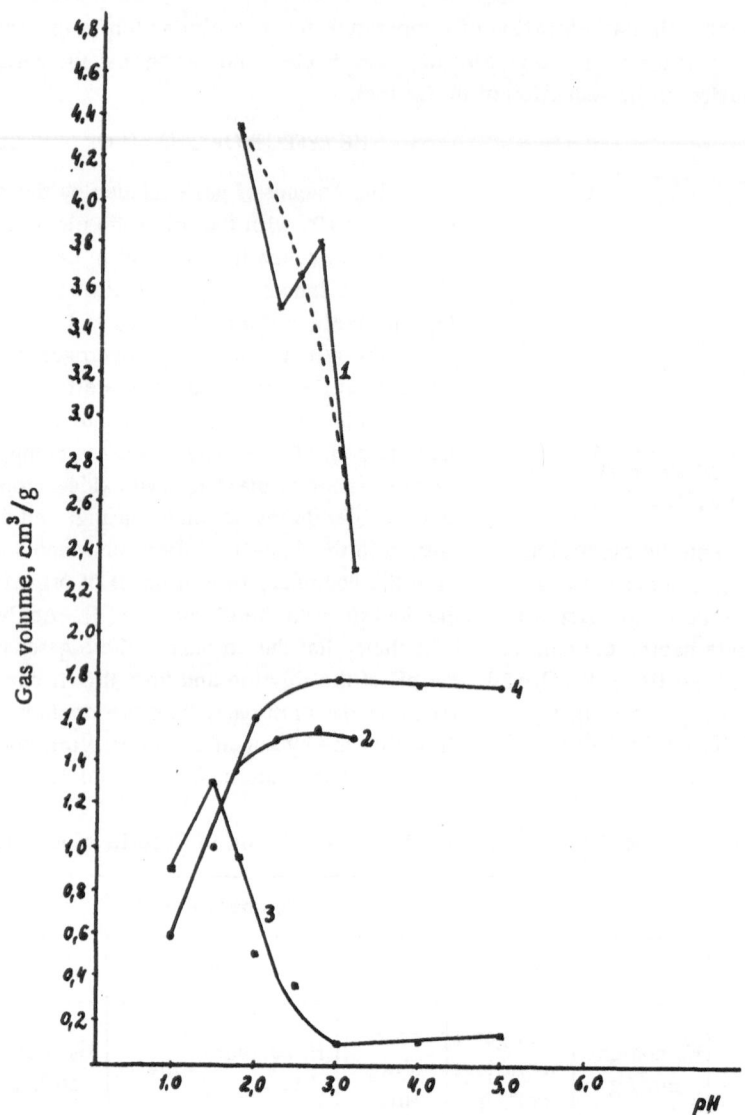

Fig. 18. The relation between the pH of the electrolyte and the amount of hydrogen present in iron and cobalt (Vagramyan, Petrova). Iron deposit obtained from 1 N $FeSO_4$ at t = 25° and I_c = 2.0 amp/dm^2. Cobalt deposit obtained from 1 N $CoCl_2$ at t = 25° and I_c = 2.0 amp/dm^2. 1) Amount of hydrogen in iron; 2) amount of gas arising from the decomposition of the hydroxides in iron; 3) amount of hydrogen in cobalt; 4) amount of gas arising from the decomposition of the hydroxides in cobalt.

Study of the relation between pH and hydrogenation in the case of electrolytic cobalt obtained from $CoCl_2$ has led to results which are similar, to a certain degree, to those found with iron from sulfate electrolytes. Curve 3 (Fig. 18) shows the variation of the hydrogen occlusion with pH and Curve 4, the variation of the occlusion of hydroxides. These curves indicate that the occlusion of hydroxides increases up to pH 3 and then remains practically constant. On the other hand, the hydrogen occlusion falls precipitously as the pH changes from 1.5 to 3·and then remains essentially unaltered as the pH is raised to 5. The small diminution in the hydrogen occlusion at pH 1.0 could be due to structural changes in the deposit and the inaccuracies in the analyses. These figures also make it clear that high occlusion of hydrogen is accompanied by very low occlusion of hydroxides, and conversely.

The hydroxide occlusion of electrolytic cobalt is considerably greater than the hydrogen occlusion, while the absolute value of the amount of hydrogen is less than in the case of iron. It is quite likely that hydrogen is less strongly bound to cobalt than to iron.

It is interesting to note that the hydrogen occlusion is less than the hydroxide occlusion in electrolytic nickel deposits obtained from acid solutions.

Data on the occlusion of hydrogen and hydroxides in the metals of the iron group are compiled in Table 10.

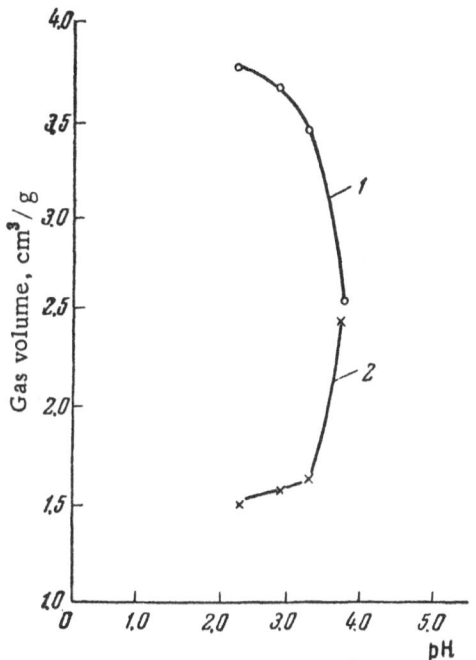

Fig. 19. The effect of aluminum sulfate on the occlusion of gases in an iron deposit (Vagramyan, Petrova). Deposit obtained from 1 N $FeSO_4$ + 100 g/liter $Al(SO_4)_3 \cdot 18H_2O$.

TABLE 10. The Content of Hydrogen and Gases from Hydroxides in Deposits of the Metals of the Iron Group Obtained from Sulfate Electrolytes at pH 1.5

Metal	Hydrogen cm^3/g	Other gases cm^3/g
Fe	4.40	1.1
Co	1.28	1.1
Ni	0.60	2.0

This table shows that the hydrogen occlusion diminishes in passing from iron to nickel while the hydroxide occlusion falls off in the opposite direction. This indicates that the hydrogen affinity on the metals of the iron group increases, and the affinity for hydroxides diminishes, in going from nickel to iron. The amount of hydrogen occluded during the electrodeposition of a metal is much greater than that required for establishing equilibrium. It is quite likely that this excessive occlusion should be ascribed to the specific details of the electrodeposition. It can be assumed that the occlusion arises principally from adsorption of hydrogen on each of the freshly deposited metallic layers. A basis for evaluation of the hydrogen occlusion is found in the work of Kavtaradze on the adsorption of hydrogen by the metals of the iron group where it has been shown that there is a 1 : 2 ratio between the numbers of adsorbed hydrogen atoms and metal atoms.

If the deposition of iron is assumed to proceed through the formation of layers 10-1000 atoms thick, the volume of hydrogen taken up by one gram of iron would be approximately 0.4 cm^3. This figure is considerably greater than the equilibrium solubility. Experimental data shows that the hydrogen occlusion varies from 0.1 to 0.5 cm^3/g in electrolytically deposited iron. From this, it can be concluded that adsorption is not the only mechanism for the occlusion of hydrogen in a deposit.

THE EFFECT OF THE STRUCTURE OF THE METAL

Nasyrova [7] has shown that the hydrogen content of carburized steel ($cm^3/100$ g) is markedly dependent on the structure:

Martensite	6.9
Troostite	15.9
Sorbite	46.5

A study of the effect of thermal treatment on the structure and hydrogenation of electrolytically plated metals by Vol'pert [8] has shown that plating processes should use tempered steel of martensite structure, since this has a somewhat distorted crystal lattice and is only slightly permiable to hydrogen. Use of carburized steel, annealed at high temperature, does not assure successful plating. Thus steel with troostite structure which has been annealed at 400° will occlude 3.8 times more hydrogen in electrolysis than a steel of martensite structure. Annealing at 600° produces a sorbite structure and increases the hydrogenation in comparison with martensite steel by a factor of 8.8.

Vol'pert has shown that spontaneous evolution of hydrogen leads to a partial (20-60%) regeneration of the mechanical properties of steel platings with the passage of time. It is the opinion of this author that it is the atomic hydrogen dissolved in the iron lattice which is most readily liberated at room temperature. Removal of the molecular hydrogen requires that heating be carried out at 400°, or higher. It is quite likely that this situation arises from the fact that atomic hydrogen can diffuse freely through the metal and is, therefore, easily lost; whereas loss of molecular hydrogen is possible only if the molecules are broken down into atoms or if cracks and defects exist in the plating. Molecular hydrogen will be retained in the metal for a rather long time, if the pressure is not high enough to cause the plating to swell.

LITERATURE

1. J. Billiter, Principles of Electroplating [Russian translation] (ONTI, Moscow, 1937).
2. F. Foerster, Elektrochemie wässeriger Lösungen (Leipzig, 1922).
3. Collection, Studies on the Corrosion of Metals under Stress [in Russian] (Mashgiz, Moscow, 1953) p. 68.
4. O. S. Popova and K. M. Gorbunova, Zhur. Fiz. Khim. 32, 2020 (1958).
5. R. Lentz, Zhur. Fiz. Khim. Obshchestvo 2, 57 (1870); Bull. de l'Acad. Imperiale des Sciences de St.-Petersbourg 14, 337 (1870).
6. A. L. Rotinyan, É. Sh. Ioffe, E. S. Kozich, and Yu. I. Yusova, Doklady Akad. Nauk SSSR 104, 753 (1955).
7. S. Nasyrova, Stal' 6, 542 (1948).
8. G. D. Vol'pert, Abstracts of Reports, Section on Metal Plating and the Chemical Treatment of Metals, All-Union Scientific-Technical Conference on the Corrosion and Protection of Metals [in Russian] (Profizdat, Moscow, 1958), Collection 4.

Chapter V

THE RATE OF EVOLUTION OF HYDROGEN AND
THE MECHANISM OF ITS OCCLUSION IN THE DEPOSIT

The strength of bonding between the metal and hydrogen will fix both the amount of gas occluded by the deposit and the rate of hydrogen ion reduction. Three types of hydrogen—metal bonding are known [1].

1. Hydrogen unites with certain metals to form saltlike compounds such as NaH and KH. Here, the hydrogen functions as a negative ion and the bonding is ionic or polar. Compounds of this kind are formed with the alkali and alkaline earth metals; they are quite stable and have high melting points and heats of formation.

2. Hydrogen unites with certain other metals to form compounds which are designated as covalent hydrides; some of these (AsH_3 and SbH_3, for example) are volatile.

3. Hydrogen unites with still other metals to form hydrides in which the electrons are mobile and the bonding metallic. Such compounds show the high thermal electrical conductivity which is characteristic of the transition metals and are highly coordinated structurally, just as the ionic compounds. They differ from the ionic compounds, however, in that the bonding involves chemically related atoms, just as in metallic alloys, rather than particles of radically different types.

It is possible that several types of atomic interaction will be set up simultaneously, depending on the nature of the metal and the presence of alloying substances which are capable of altering the bond character.

It is clear that the rate of hydrogen ion reduction must be affected by the strength of bonding between the metal and hydrogen. The depolarizing effect of the cathode metal on the rate of hydrogen ion reduction can be observed in the evolution of hydrogen from metals such as arsenic and antimony which form gaseous hydrides of high volatility.

Several theories have been proposed to account for the observed differences in the rate of hydrogen ion reduction on various metals. These theories of hydrogen overvoltage are based on the assumption that hydrogen ion reduction takes place in successive steps (dehydration, discharge, recombination, bubble formation, etc.) and that one or more of these steps is retarded. Thus, the rate of reduction can be fixed by the rate of ionic discharge [23], or by the rate of recombination of atoms to molecules on the metal surface [4], or by the rate of the so-called electrochemical desorption [5].

The theory of retarded discharge considers that the reduction of hydrogen ions to adsorbed atoms,

$$H^+ + e \rightarrow H_{ads}$$

is a reaction of finite velocity whose rate is fixed by the difference of potential between the metal and the solution. The original formulation of this theory assumed the retardation in the evolution of hydrogen to be related to electrochemical discharge [2, 3]. More recently, the protagonists of this theory incline to the view that the union of the positive hydrogen ion with a negative electron could scarcely involve an appreciable activation energy and they, therefore, consider the retardation of reduction to be due to the strong hydration of the hydrogen ion in solution:

$$H_3O^+ + e \rightarrow H_{ads} + H_2O.$$

It then follows that the discharge rate would be related to the electrode potential by an equation of the form:

$$I_K = k \cdot [H^+] e^{-\frac{F}{RT}(\alpha\varphi - \beta\psi_1)},$$

This equation indicates that an increase in the cathode potential should lead to a rise in the current density and an increase in the rate of hydrogen ion reduction. It also follows that there should be a linear relation between the electrode potential and the logarithm of the current density. The experimental data conform, rather well, to this equation. No clear cut treatment of the problem of the dependence of the rate of hydrogen ion reduction on the nature of the cathode metal can be had from this theory.

The recombination theory of the hydrogen overvoltage considers the effect of the cathode metal on the rate of hydrogen ion reduction [4].

Here, the retardation of hydrogen ion reduction is assumed to be due to the slow removal of atomic hydrogen from the electrode surface. This leads to an accumulation of excessive quantities of adsorbed atomic hydrogen on the electrode and results in a displacement of the electrode potential in the negative direction by an amount which is to be calculated through the equation:

$$\varphi = k + \frac{RT}{nF} \ln \frac{[H^+]}{[H_{ads}]} .$$

The rate of recombination is fixed by the nature of the metal and its surface state; therefore, it follows that the rate of loss of atomic hydrogen from the surface will vary from case to case. This is sometimes designated as the "catalytic" mechanism of hydrogen evolution. It is an interesting fact that there is a parallelism between the rate of electrochemical reduction of hydrogen on various metals and the catalytic activity of these metals with respect to atomic hydrogen. A listing of the metals in order of increasing hydrogen overvoltage is, at the same time, a listing in order to diminishing catalytic activity, i.e.,

Catalytic activity
←

Pt, Pd, W, Ni, Fe, Ag, Cu, Zn, Sn, Pb
⟶

Hydrogen overvoltage

This indicates the existence of a relation between the retardation of hydrogen ion reduction and the ability of a metal to catalyze the recombination of hydrogen atoms.

The above sequence also lists the metals in order of their ability to adsorb hydrogen, the hydrogen overvoltage increasing, and the hydrogen adsorption diminishing, in passing from platinum to lead.

Kobozev and Nekrasov [4] have derived the following equation in which the hydrogen overvoltage is related to the nature of the metal through the adsorption energy, W_{ads}:

$$\eta = k - \frac{W_{ads}}{F} + \frac{RT}{F} \ln [H]_S .$$

This equation shows that the overvoltage diminishes as the energy of hydrogen adsorption increases. This theory is in good qualitative agreement with experiment and gives an interpretation of the observed relation between the nature of the metal and the rate of hydrogen ion reduction.

The principal objection to the recombination theory is that the coefficients in the above equation do not agree with the experimental data. Temkin [6] has shown, however, that this defect can be eliminated by taking account of the nonuniformity of the electrode surface.

There is also a theory which rests on the so-called electrochemical desorption of adsorbed hydrogen [6]:

$$H_{ads} + H^+ + e \rightarrow H_2.$$

Here the reaction rate is determined by the concentration of atomic hydrogen on the electrode surface. This concentration is extremely low on some metals (such as Hg and Pb) and the retardation of reduction is presumed to be due to this fact. There are a number of cases in which the effect of the nature of the metal on the rate of hydrogen ion reduction can also be interpreted in terms of this theory.

It should be noted that the adherents of these various mechanisms for the evolution of hydrogen frequently set one point of view in sharp opposition to another. Frumkin [7] has justly observed that "the one factor or the other will effect the reaction kinetics in large measure, depending on the nature of the electrode and the experimental conditions," i.e., "retardation of the loss of molecular hydrogen and retardation of discharge must both be taken into account."

TABLE 11. Hydrogen Overvoltage (volts) in 2 N H_2SO_4.

Current density, amp/cm²	Re	Pd	Au	W	Mo	Ni	Co	Ag	Cr	Fe	Ge	Ta
10^{-3}	0.12	0.14	0.24	0.26	0.35	0.30	0.32	0.35	–	0.37	0.39	0.46
10^{-2}	–	–	0.39	0.37	0.50	0.50	0.39	0.43	–	0.46	0.62	0.55
10^{-1}	–	–	0.59	0.46	0.56	0.48	0.42	0.50	0.67	0.60	0.85	0.61

Current density, amp/cm²	Cu	Cd	Sn	Al	Be	Nb	Bi	In	Zn	Hg	Tl	Pb
10^{-3}	0.48	0.51	0.57	0.58	0.63	0.65	0.69	0.80	0.83	1.04	1.05	1.12
10^{-2}	0.60	0.71	0.71	0.68	–	–	0.83	–	1.01	1.15	–	1.24
10^{-1}	0.74	0.93	0.83	0.76	–	–	0.91	–	1.17	1.21	–	1.26

Certain experimental data on the effect of the metal on the hydrogen·overvoltage [8] will not be considered (Table 11).

The table shows the hydrogen overvoltage and the rate of reduction to increase in the order: Re, Pd, Au, W, Mo, Ni, Co, Ag, Cr, Fe, Ge, Ta, Cu, Cd, Sn, Al, Be, Nb, Bi, Th, Zn, Hg, Tl, Pb.

Theory indicates that metals with low hydrogen overvoltage should adsorb a great deal more atomic hydrogen then metals with high overvoltage. Thus, there must be a parallelism between the hydrogen overvoltage and the degree of hydrogenation [9]. A comparison of the above ordering of the metals based on hydrogen overvoltage with data on hydrogenation shows that the strongly hydrogenated metals have low overvoltages, and vice versa. It is possible that certain departures from this general rule are due to the fact that additional factors, such as the presence of alloying substances or the adsorption of surface active substances, can increase or diminish the hydrogen adsorption and thereby veil the dependence of hydrogenation on the nature of the metal. There are, on the other hand, cases such as that of palladium, in which rapid loss of hydrogen from the metal makes for difficulty in the determination of hydrogenation. The result is that it is principally the equilibrium fraction of hydrogen which enters into some comparisons of hydrogenation, whereas, the equilibrium plus the nonequilibrium fraction is considered in others. The occlusion of hydrogen during the electrolytic separation of a metal can proceed either by mechanical pickup or by adsorption of atomic hydrogen on the surface. A part of the adsorbed hydrogen enters into the lattice where it forms a solid solution or intermetallic compound, while the remainder recombines to molecular hydrogen. The hydrogen ions discharge simultaneously with the metallic ions and adsorb on the crystal faces where they are occluded as the lattice grows, occupying lattice points or distributing themselves in the interstices between them. It is possible that the formation of a solid solution does not involve adsorption of atoms into the lattice, but direct incorporation of protons which are counterbalanced by the metal electrons. The possibility of the protonic type of hydrogen occlusion in the deposit is supported by the fact that the degree of hydrogenation is considerably greater in acid solutions than in alkali solutions. It is well known that the mechanisms of hydrogen evolution are quite different in acid and in alkaline solutions [10]. In acid solutions the reduction proceeds according to the following reaction:

$$H^+ + e \rightarrow H_{ads},$$

while the reaction in alkali solutions is probably:

$$H_2O + e \rightarrow H_{ads} + OH^-.$$

The laws of electrochemical kinetics indicate that the rate of this last reaction would be determined by:

$$I_c = k\,[H_2O]\,e^{-\frac{\alpha F\,(\varphi - \varphi_1)}{RT}}.$$

This equation shows that both hydrogen ions and neutral water molecules are involved in the reduction reaction. It is clear that the probability of protonic occlusion will be reduced in this case and the low degree of hydrogenation of metals in alkaline solutions is probably due to this fact.

Hydrogen can be picked up mechanically during the growth and adhesion of separate crystals and even gaseous hydrogen can be occluded in this way.

The mechanical pickup of hydrogen bears no relation to either the nature of the metal or the reduction rate, but is largely fixed by the character of the deposit and the conditions of its formation. Thus, these factors can also lead to a distortion in the interrelation of hydrogen overvoltage, catalytic activity, adsorption capacity, and solubility of hydrogen in the metal.

It follows from what has been said that hydrogen can be occluded in a metal by various processes and in various forms, depending on the nature of the metal, the experimental conditions, the presence of alloying substances, and the state and structure of the surface.

LITERATURE

1. D. T. Hurd, An Introduction to the Chemistry of the Hydrides (N. Y.-London, 1952).
2. T. Erdey-Gruz and M. Volmer, Z. phys. Chem. 150, 203 (1930).
3. A. N. Frumkin, Z. phys. Chem. A, 160, 116 (1932).
4. N. I. Kobozev and N. I. Nekrasov, Z. Elektrochem. 36, 529 (1930).
5. J. Heyrovsky, Rec. Trav. Chim.-Pays-Bas 44, 499 (1925).
6. M. I. Temkin, Zhur. Fiz. Khim. 15, 296 (1941).
7. A. N. Frumkin, V. S. Bagotskii, Z. A. Iofa, and B. N. Kabanov, The Kinetics of Electrode Processes [in Russian] (Moscow State University Press, 1952).
8. A. G. Percherskaga and V. V. Stender, Zhur. Fiz. Khim. 24, 856 (1950); Zhur. Priklad. Khim. 19, 1303 (1946); A. Hickling and F. Salt, Trans. Faraday Soc. 36, 1226 (1940); 37, 224, 333, 450 (1941); J. O'M. Bockris, Trans. Faraday Soc. 43, 417 (1947).
9. M. Smyalovskii, Byull. Pol'sk. Akad. Nauk. Otdel 2, 4, 1 (1957).
10. Yu. V. Baimakov and M. I. Zamotorin, Proceedings, Conference on Electrochemistry [in Russian] (Akad. Nauk SSSR Acad. Sci. USSR Press, 1953) p. 125.

Chapter VI

THE EFFECT OF HYDROGENATION ON THE OVERVOLTAGE FOR METAL DEPOSITION

The problem of relating hydrogenation and overvoltage for metal deposition is just as significant as that of fixing the companion relation between overvoltage and hydrogen occlusion. For this reason, study will not be made of certain data, on the electrodeposition of iron, which have been obtained by Kuznetsova. Figure 20 shows the relation between the pH of the electrolyte and the overvoltage (Curve 1) and current efficiency (Curve 2) for iron deposition. The figure brings out the fact that the current efficiency rises sharply (from 10 to 90%) over the pH interval 1.5-2.5 and then remains essentially constant as the pH is increased further.* There is, at the same time, a maximum at pH 2.0-2.5 on the curves relating the electrode potential and pH. Comparison of these curves shows the maximum on Curve 1 to correspond to a current efficiency of 50-60%. From this it follows that hydrogen reduction accounts for the major proportion of the current consumed along the ab segment of Curve 1, this proportion increasing in passing from b to a. It is the deposition of metal which consumes the greater proportion of the current along the bc segment of this curve and the fraction of current so consumed, rises in passing from point b to point c.

Thus, at fixed rate of reduction the hydrogen ion discharge predominates in the range of higher acidities, while metal ion discharge is favored over the segment bc where hydrogen ion reduction is retarded strongly. The retardation of metal ion reduction on a hydrogenated electrode can be interpreted by supposing that hydrogen penetrates into the crystal lattice of a metal of the iron group to form a compound of the type $Me^--H_2^+$, with the positive charge oriented toward the electrolyte. The presence of this positive charge on the electrode surface serves as an added impediment to the reduction of metallic ions.

This interrelationship is brought out again in a comparison of the curves of Fig. 18, the one showing the variation of the polarization, η, for iron deposition with the pH of the electrolyte and the other, the dependence of the amount of occluded hydrogen, v, on this same parameter. These curves point up the fact that the greatest occlusion of hydrogen occurs at the point corresponding to the maximum overvoltage in the investigated pH range; they also show that an increase in the pH of the electrolyte will reduce both the overvoltage and the degree of hydrogenation of the iron deposit. Comparison also indicates that the retardation of metal ion reduction increases in parallel with the amount of hydrogen occluded in the deposit.

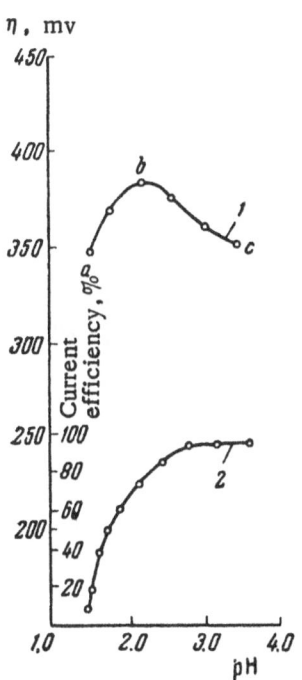

Fig. 20. The relation between the pH of the electrolyte, the overvoltage (1), and the current efficiency (2) for deposition of iron (Vagramyan, Kuznetsova, et al.). Iron deposited from 1 N $FeSO_4$ at t = 25° and I_c = 2.2 amp/dm^2

---

*It should be noted that this is exactly the pH interval within which dense, homogeneous, shiny iron deposits are obtained.

31

It is obvious that the η versus pH curve falls off more gradually than the v versus pH curve, the indication being that hydroxides also affect the retardation of the metal ion reduction, although to a smaller extent than the occluded hydrogen.

Comparison shows that an elevation of the temperature from 25 to 75° invariably diminishes both the gas occlusion in, and the overvoltage for, electrodeposition of the metals of the iron group (Fe, Ni, Co) (see Table 9). The figures of the table are in good agreement with other results which have shown the rate of electrochemical reaction to be vitally affected by the state of the electrode surface. It is clear that the surface state is determined largely by the number of foreign particles which are occluded in the electrolytic deposit. The presence of such particles on the electrode surface impedes the reduction of the discharging ions and gives rise to an overvoltage of considerable magnitude. This accounts for the observed parallelism between the overvoltage for metal deposition and the occlusion of hydrogen and other gases.

TABLE 12. The Relation between Internal Stress, Hydrogenation, and Overvoltage in Metals

Metal	Current efficiency, %	Electrolytic composition, g/liter	Current density, amp/dm^2	Volume of gas in deposit, cm^3/g	Internal stress, relative units	Over-voltage, mv
Copper	100	$CuSO_4$ – 250 H_2SO_4 – 50	3 - 1.0	0.036 – 0.086	+0.9 – 1.65	50 –168
Cobalt	98	$CoSO_4$ – 1 N H_3BO_3 – 30	2.0	0.86 – (2.3)	5.8	400
Nickel	98	$NiSO_4$ – 210 H_3BO_3 – 30 KCl – 10	2.0 - 3.0	0.4 - 1.07 (2.5)	4.7 - 6.5	500- 550
Copper	100	$CuSO_4$ – 250; H_2SO_4 – 50 $CuSO_4$ + 0.005 thiourea $CuSO_4$ + 0.5 naphthalenedisulfonic acid	30	0.086 0.5 0.18	1.45 3.6 10.65	168 250 206

An analysis of the experimental data on metallic hydrogenation and overvoltage shows the following general relations to exist between these factors.

There is practically no occlusion of hydrogen in cadmium, zinc, tin, lead, copper, silver, and other metals which deposit out at essentially zero overvoltage. The overvoltage is generally high in the deposition of nickel, cobalt, chromium, manganese, and other metals which take up considerable quantities of hydrogen during electrolysis. And, finally, very great quantities of hydrogen are dissolved by niobium, thorium, vanadium, zirconium, and other metals which cannot be deposited from aqueous solution.

These facts indicate that the adsorption and occlusion of hydrogen in the deposit may actually be the principal factors in fixing the retardation of metal ion reduction. It should be noted that a quantitative comparison of the effect of these factors is difficult, since the action of the occluded hydrogen varies with the mechanism of occlusion.

Comparison shows that the overvoltage, the number of occluded foreign particles, and the internal stress in the electrolytic deposit are all interrelated. The existence of such relationship is observed in studies on the deposition of a single metal under various conditions, as well as in comparative studies on different metals.

Our own experimental data on the electrodeposition of various metals are collected in Table 12.

This table makes it clear that the internal stress and the overvoltage both rise with an increase in the number of foreign particles occluded in the deposit.

Chapter VII

THE ADSORPTION OF SURFACE-ACTIVE
SUBSTANCES IN ELECTRODEPOSITION

It has been noted above that adsorption is the principal mechanism for the occlusion of surface-active substances in the body of a deposit. A metal can absorb small particles such as the hydrogen atom, but there is practically no possibility of a similar occlusion of rather large organic molecules. Thus surface-active substances must be taken into a deposit by adsorption on expanding crystal faces with subsequent occlusion during crystal growth.

It is usually considered that electrode adsorption occurs over a definite potential interval which is fixed by the null charge point, anions adsorbing essentially on a positively charged surface, cations on a negatively charged surface, and neutral organic molecules on a surface which is at the null charge potential. There is no possibility of the adsorption of neutral molecules and organic ions on a highly charged surface.

These ideas have come out of studies on the adsorption of surface-active substances on mercury and it is doubtful that they are fully applicable to a discussion of adsorption on solid metals.

Both physical (van der Waals) and chemical adsorption must be considered in discussing adsorption on metallic surfaces. The type of adsorption is largely fixed by the nature of the metal and the state of its surface. The metallic surface is generally nonuniform, the potential energy varying from point to point. Preferential adsorption will occur on one sector of the surface or another depending on the type of strength of bonding between the metal and the adsorbate.

Surface diffusion in solids, i.e., the migration of surface atoms or molecules resulting from thermal agitation, was first detected by Volmer and Estermann. Since then, the existence of this effect has been confirmed in various studies.

Migration of adsorbed particles over the metallic surface (nonlocalized adsorption) will diminish the likelihood of a surface-active substance entering the deposit, since it will give rise to a mixing of these particles with the expanding metal layers as electrocrystallization proceeds. Furthermore, the retardation of metal ion reduction from mobile adsorbed particles will be much less than that arising from particles which are fixed in position. Thus, the degree of retardation of ion discharge will vary with the type of adsorbed particles.

The nature of the adsorbent is a prime factor in fixing the magnitude of chemical adsorption. Decomposition may accompany the adsorption of certain types of organic molecules on certain metals. An instance of this kind found in the adsorption of ethylene on a transition metal. Here, hydrogen is partitioned according to the equation:

$$4Me + C_2H_4 \rightarrow Me_2C_2H_2 + 2MeH$$

and the chemically adsorbed hydrogen then reacts with the ethylene to form ethane:

$$2MeH + C_2H_4 \rightarrow 2Me + C_2H_6.$$

The hydrogenation of various organic substances proceeds through a mechanism of this type.

Irreversible adsorption also takes place on various component parts during the electrodeposition of a metal and results in the occlusion of both the substance and its decomposition products in the deposit. The crystallographic structure of the electrolytic deposit is quite complex. Such a deposit is built up from crystals of various forms and dimensions, variously oriented with respect to one another and it, therefore, contains voids, microchannels, cracks, and other defects. The state of a surface-active substance, or a decomposition product resulting from such substance, will depend on the position which it occupies in the deposit after inclusion. It is clear that large molecules can be occluded in the voids and crystal lattice defects, whereas only ions and atoms can be taken into the lattice itself.

The rate of adsorption is one of the basic factors in a treatment of the electrodeposition of metals, since the crystal surface is being renewed continuously during deposition and the number of occluded foreign particles is fixed by the ratio of the rates of deposition and adsorption of surface-active materials.

The Laboratory for the Electrodeposition of Metals has developed methods for determining the rate of adsorption of surface-active substances which are based on the assumption that such adsorption passivates the metallic surface. The rate of passivation is then a measure of the adsorption rate.

Study of the rate of passivation of the electrode involves the following principles [1]. When the current is interrupted and the cathode allowed to stand in the solution, it is observed on reclosing the circuit that a part of the surface has been passivated. The initial cathodic polarization ($\eta_{t=0}$) is excessively high, and only gradually falls off to a stationary value ($\eta_{t=\infty}$) as deposition of metal and surface activation proceed (Fig. 23). The more extensive the interruption of the electrolysis (τ) and the longer the electrode is allowed to stand in the solution, the higher the instantaneous cathodic polarization on reclosing the circuit (in comparison with the stationary value, potential rise $\Delta\eta$). Here $\Delta\eta = \varphi(\tau)$, with

$$\Delta\eta = \eta_{t=0} - \eta_{t=\infty}.$$

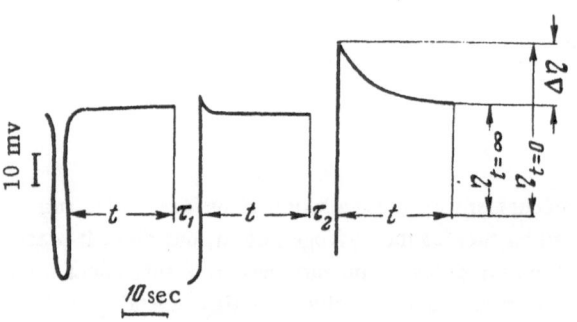

Fig. 21. The time variation of the cathodic polarization before, and after, closing the circuit.

This effect becomes quite pronounced when certain surface-active substances are introduced into the solution. Observations show that fewer nuclei are formed on the cathode surface after reclosing the circuit than were being formed before the circuit was broken [2]. Thus, the electrode surface undergoes passivation while standing in the solution and the rate of metal deposition is reduced, thereby. It is quite likely that this passivation of the surface is the cause of the excessive electrode polarization which is observed when the circuit is closed once more. It is, in fact, true that a reduction in the active surface area by adsorption of foreign molecules will lead to an increase in the actual current density and the electrode polarization. Both values will, however, diminish as deposition of metal proceeds and the active surface of the electrode rises, and each will approach a stationary limit corresponding to equality between rates of passivation and deposition and constancy of active surface area [3].

Thus, the amount of adsorption of a surface-active substance on an electrode can be estimated by measuring the polarization on reclosing the circuit after current flow has been interrupted. The fraction of the electrode surface covered with molecules of the surface-active substance is proportional to the increase in the polarization ($\Delta\eta$) which is observed on resumption of electrolysis. The variation of $\Delta\eta$ with the length of time during which electrolysis is interrupted, i.e., the slope of the $\Delta\eta = f(\tau)$ curve, must be known in order to evaluate the rate of adsorption of the surface-active substance.

The rate of adsorption of the surface-active substance is found in the following manner. The covered fraction of the electrode surface is calculated by assuming that the neutral organic molecules form monomolecular adsorption layers. Such calculation requires that the true current density at the instant of resuming electrolysis be obtained from the polarization at this instant, $\eta_{t=0}$, and an $\eta = \varphi(I)$ polarization curve which has been developed for the solution without the surface-active substance, using a rapid method and working at current densities high enough to assure full activity of the electrode surface. The diminution of the active surface, ΔS, during interruption of electrolysis is then found by comparing the true current density corresponding to the observed polarization, $\eta_{t=0}$ and the current density corresponding to the stationary value. The $\Delta S = \varphi(\tau)$ equation is developed by carrying out a series of experiments in which the length of time of interruption is varied; this fixes the rate at which the electrode surface is covered by adsorbed molecules, or the rate of adsorption.

It should be kept in mind, however, that the precision of this method can be adversely affected by concentration polarization. The concentration polarization will be considerably less at the instant of reclosing the circuit than it was before the current was interrupted, the concentration of discharging metal ions in the layer immediately surrounding the cathode being reduced during electrolysis and built up again while the current is turned off. The concentration polarization arising from this build-up in population of discharging ions at the electrode surface can reduce, or even overcompensate, the rise in polarization which is due to passivation of the cathode surface during interruption of the current. It follows that the sensitivity of this method is improved by eliminating concentration polarization. For this reason studies of rates of electrode passivation are carried out in concentrated solutions, using cathodes of small surface area and working at low current densities.

Rates of electrode passivation are measured with a special apparatus which has been described in detail in [1].

Certain data on the passivation of silver, zinc, and nickel and the rates of adsorption of surface-active substances during electrodeposition of these metals will be considered now.

THE ADSORPTION OF SURFACE-ACTIVE SUBSTANCES IN THE ELECTRODEPOSITION OF SILVER FROM NITRATE SOLUTIONS

The electrolytes from which metals are deposited will always contain organic contaminants and it frequently happens that these cannot be eliminated completely, regardless of the care used in purification. Thus, the adsorption of foreign particles on the cathode is not an extraneous effect but a factor which will invariably influence the character of an electrolytic deposit which has been obtained under ordinary conditions. In certain cases, adsorbing substances are produced in electrolysis, an instance being the formation of hydroxides during the electrodeposition of metals of the iron group [4].

There is a potential jump on the passivation curve for a silver cathode which is undergoing electrodeposition of silver from a solution of chemically pure $AgNO_3$ in doubly distilled water that is free of added surface-active substances [4]. Such potential jump on resuming electrolysis after current interruption is indication that some surface-active substance is being adsorbed in an amount which increases with time. Only a very careful purification of the working solution of $AgNO_3$ with oxidation of the organic contaminants by platinized platinum in an atmosphere of oxygen will eliminate this rise in polarization following temporary interruption of electrolysis. Such purification can be carried out only under laboratory conditions. These facts indicate that electrodeposition of metal on the surface of a silver cathode will always be accompanied by adsorption of those traces of organic substances which are invariably present in solutions that have not been specially purified. Rate curves for the passivation of a silver cathode in three different $AgNO_3$ solutions, one unpurified, one carefully purified, and the third with added dextrine are given in Fig. 22. The rates of adsorption of surface-active substances in the three solutions are measured by the slopes of the curves. Curve 1 shows that there is absolutely no adsorption of surface-active substances in the purified solution, $\Delta\eta$ being equal to zero even on resuming electrolysis after a 15 min current interruption. The organic substances present in the unpurified $AgNO_3$ solution adsorb out on the cathode at a definite rate (Curve 2). Both the rate and absolute magnitude of adsorption from this solution are increased by the addition of 0.2% dextrine (Curve 3).

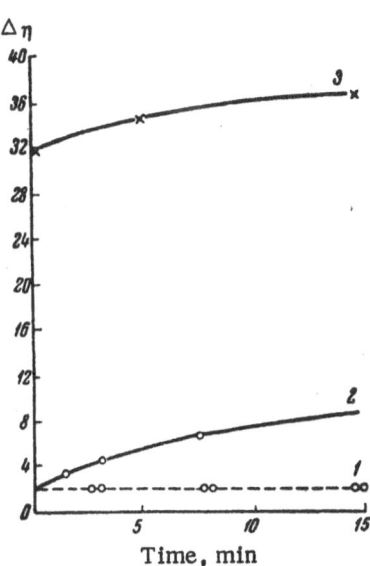

Fig. 22. The rate of passivation of a silver cathode in $AgNO_3$ solutions of various degrees of purity (Vagramyan): 1) Carefully purified solution; 2) unpurified solution; 3) solution with 0.2% added dextrine.

Although a solution will almost invariably contain substances which are readily adsorbed on the cathode, there are other organic materials which will not adsorb on silver. Thus the form of the passivation curve and the magnitude of the polarization in a carefully purified $AgNO_3$ solution are not in the least altered by the addition of saturated, normal alcohols. This is indication that these alcohols are not adsorbed on the electrode surface and exert no influence on the electrode reaction.

THE ADSORPTION OF ALCOHOLS IN THE ELECTRODEPOSITION OF ZINC

Vagramyan and Solov'eva [1] have studied the adsorption of certain saturated, normal alcohols (amyl, hexyl, and octyl) in the electrodeposition of zinc, and the relation of this adsorption to the concentration of the alcohol and the length of its hydrocarbon chain.

Figure 23 represents the passivation curve obtained at $I_c = 2$ ma/cm^2 in a purified 2 N $ZnSO_4$, free of added surface-active substances. The fact that there was no increase in polarization following interruption of electrolysis is indication of the high purity of the $ZnSO_4$ solution which was used here.

The form of the passivation curve in this solution is altered markedly by the introduction of a small amount of octyl alcohol(0.0115 mole/ liter). Figure 24 shows that the presence of the alcohol gives rise to a pronounced potential jump following interruption of electrolysis, the value of $\Delta \eta$ increasing perceptibly with the length of the interruption period to give a stationary polarization which is much higher than in the pure $ZnSO_4$ solution. A passivation curve of this type is evidence of a slow adsorption of octyl alcohol on the surface of the zinc electrode. The fraction of the surface covered by molecules of the surface-active substance can be evaluated from the passivation curve and the polarization curve developed for the electrodeposition of zinc from the pure electrolyte. The curves of Fig. 25 have been used as a basis for calculations of this kind and have led to the results presented in Table 13.

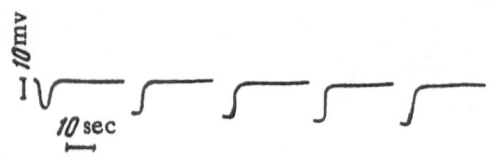

Fig. 23. passivation curve for a zinc cathode in 2 N $ZnSO_4$; I_c = 2 ma/ cm² and t = 18° (Vagramyan, Solov'eva).

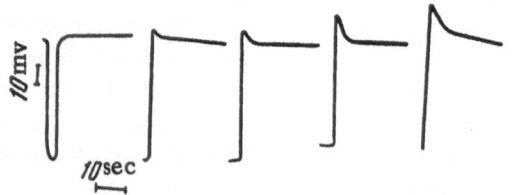

Fig. 24. Passivation curve for a zinc cathode in 2 N $ZnSO_4$ containing added octyl alcohol; I_c = 2 ma/ cm² and t = 18° (Vagramyan, Solov'eva).

Figure 25 shows the time variation of the fractional coverage of the surface of a zinc electrode by octyl alcohol molecules. The slope of any one of these curves is a measure of the rate of adsorption of octyl alcohol on zinc. This figure makes it clear that the rate of adsorption is initially high and then falls off. The fact that the magnitude and the rate of adsorption both increase with increasing concentration of alcohol in solution is brought out by a comparison of Curves 1-3 of this figure. The values of $\Delta \eta_{t=0}$ and the surface coverage are constant for interruption periods (τ) of 5 min or more, so that the surface is apparently saturated under these conditions. An isotherm for the adsorption of octyl alcohol can be obtained

TABLE 13. Calculation of the Adsorption of Octyl Alcohol from an Alcohol Saturated 2 N $ZrSO_4$ Solution at 18°

$\eta_{t=0}$, mv	True current density, ma/ cm²	Increase in current density, I'/ I	Active surface area, S_{act}, cm²	Diminution of active surface area, ΔS, cm²	Surface coverage, %
62.8	42.75	5.25	$0.365 \cdot 10^{-3}$	$1.55 \cdot 10^{-3}$	80.9
75.2	52.50	6.45	$0.297 \cdot 10^{-3}$	$1.62 \cdot 10^{-3}$	84.8
92.5	64.50	7.93	$0.242 \cdot 10^{-3}$	$1.67 \cdot 10^{-3}$	87.8
101.0	71.25	8.76	$0.219 \cdot 10^{-3}$	$1.70 \cdot 10^{-3}$	88.6
108.0	76.50	9.49	$0.202 \cdot 10^{-3}$	$1.71 \cdot 10^{-3}$	89.4

from these curves (Fig. 26). Similar calculations have been made on the adsorption of hexyl and amyl alcohols. Here, too, the magnitude of adsorption quickly approaches a constant value in saturated solutions, whereas surface saturation is reached much more slowly in dilute solutions.

The adsorption of these alcohols increases with increasing length of the hydrocarbon chain, even though comparison must be made against solutions containing lower concentrations of octyl alcohol than hexyl or amyl, the low solubility of octyl alcohol rendering impossible a comparison at equal concentrations.

The adsorbability of these alcohols from saturated solutions is almost the same in each case (it is clear that surface saturation is attained), although the adsorption rates are different.

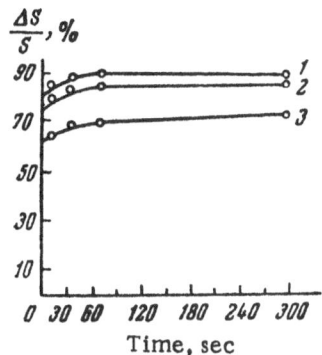

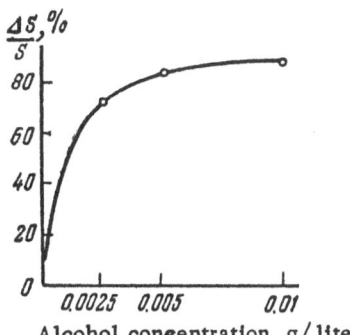

Fig. 25. Curves showing the rate of
adsorption of octyl alcohol on zinc in
2 N ZnSO$_4$ (Vagramyan, Solov'eva):
1) 0.0115 M; 2) 0.0057 M; 3) 0.0027 M.

Fig. 26. Isotherm for the adsorption of octyl alcohol from
2 N ZnSO$_4$ (Vagramyan,
Solov'eva).

THE EFFECT OF THE CONDITIONS OF ELECTROLYSIS AND THE NATURE OF THE
SURFACE-ACTIVE SUBSTANCE ON THE RATE OF ADSORPTION
OF THE LATTER ON THE ELECTRODE SURFACE

Study has brought out the interesting fact that the rate of adsorption of alcohols on the surface of a zinc cathode is strongly dependent on the acidity of the zinc sulfate solution.

Figure 27 presents a passivation curve for 2 N ZnSO$_4$ (pH 4.5) saturated with octyl alcohol. The polarization
shown by this curve (90 mv) is considerably higher than that which is met in a solution free of alcohol, the indication being that the discharge of zinc ions is strongly impeded. At the same time, the polarization is seen to be no
higher after resuming electrolysis than it was prior to interruption, but rises gradually. Moreover, the polarization
is almost independent of the time during which the cathode stood in the electrolyte with the current turned off.
This is, clearly, a case in which the retarding action of the surface-active substance is independent of the time, and
the adsorption rate must, therefore, be rather high.* Thus the rate of adsorption of a surface-active substance on zinc
varies with the pH of the ZnSO$_4$ solution. This same effect is also observed in a study of passivation curves for electrodeposition of nickel from sulfate solutions containing the sodium salt of naphthalenedisulfonic acid (2,6 and 2,7
isomers). In the presence of this salt, there is no alteration in the polarization of the nickel cathode following interruption of electrolysis, possibly because the salt is adsorbed on the cathode surface at a very high rate.

The addition of thiourea brings about an abrupt rise in potential following interruption of electrolysis, the
magnitude of this rise being independent of the duration of the interruption. It is possible that this is indication of
a rather rapid adsorption of thiourea on nickel, although the adsorption occurring here must be slower than that which
is met with naphthalenedisulfonic acid. Similar results have been obtained in studies on the adsorption of surface-active substances from sulfate solutions onto copper. Still other forms of passivation curves have been obtained in a
study of the adsorption of various surface-active substances on silver electrodes, the indication being that the rate of
adsorption is not the same as in the other cases. It has been pointed out already that a finite time is required for the
adsorption of dextrine from nitrate solutions. The adsorption of butyric acid on silver is a very rapid process and
there is no potential jump following interruption of electrolysis, although the polarization is much higher than in
pure AgNO$_3$.

The two types of passivation curves which have been described here are extremes; intermediate forms can be
obtained by such devices as using mixed additives.

The literature contains numerous papers on the problem of establishing a relation between the structure of the
molecules which are adsorbed on the electrode and their affect on the rate of the electrode reaction and the structure of the electrodeposit [5]. The discussion of such work lies outside of the scope of this text.

Finally, it should be noted that the rate of occlusion and the amount of material taken up by the deposit will
determine not only the rate of the electrode reaction and the structure of the deposit, but also the physicomechanical properties of the latter. This point will be brought out in what follows.

---

*Complex formation could also account for a passivation curve of this kind, but the low concentrations of the added
substances make it unlikely that such an effect is involved here.

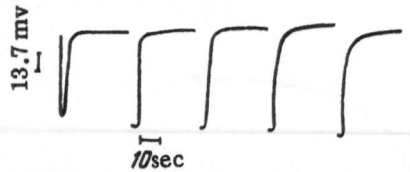

Fig. 27. Passivation curve for a zinc electrode in 2N $ZnSO_4$ (pH 4.5) at I_c = 2 ma/cm^2, t = 18° (Vagramyan, Solov'eva).

One other detail comes out of a study of the effect of foreign substances on the physicomechanical properties of the electrolytic deposit, namely, that these properties vary with the molecular stability of the surface-active substance at fixed rate of adsorption. Thus different effects can be expected depending on whether occlusion involves the substance itself or its decomposition products.

LITERATURE

1. A. T. Vagramyan and Z. A. Solov'eva, Zhur. Fiz. Khim. 24, 1252 (1950).
2. A. T. Vagramyan, Zhur. Fiz. Khim. 14, 1132 (1940).
3. A. G. Samartsev, Doklady Akad. Nauk SSSR 22, 243 (1939).
4. A. T. Vagramyan, Doklady Akad. Nauk SSSR 27, 805 (1940).
5. E. Raub and M. Wittum, Z. Elektrochem. 46, 71 (1940).
6. H. Fischer, Elektrolytische Abscheidung und Elektrokristallisation von Metallen (Berlin. Springer. Verlag, 1954).

Chapter VIII

METHODS FOR MEASURING AND EVALUATING INTERNAL STRESSES

It has been noted above that distortion of the crystal lattice and occlusion of foreign substances produce a strain in the deposit and, at the same time, give rise to forces which tend to compress or distend the deposit and return it to its normal state.

A cathode which is subject to deposition of electrolytic metal from one side only will bend under the action of these forces. There are two types of electrode bending, the type actually observed being fixed by the experimental conditions:

1) Bending in the direction of the anode. This type of bending is met when the deposit is distended and tends to reduce its volume. The deposit is then said to be in tension and the stress carries a positive sign (Fig. 28a).

2) Bending in the opposite direction, away from the anode. This type of bending occurs when the deposit is contracted and tends to increase in volume. The deposit is then under compression and the stress carries a negative sign (Fig. 28b).

Fig. 28. Stress system in the deposit: a) Tension; b) compression.

The numerous methods which have been developed for measuring internal stresses can be divided into four principal groups:

1) method of deformation of a glass bulb,
2) method of cathode bending,
3) strain gauge method,
4) x-ray method.

METHOD OF DEFORMATION OF A GLASS BULB (MILLS METHOD)

The use of a deformable glass bulb for determining the internal stress within a deposit was first reported by Mills [1]. In essence, Mills first covered the external surface of a bulb containing mercury with chemically deposited silver and then deposited a layer of the test metal on this silver, electrically. This bulb was equipped with a capillary and resembled an ordinary thermometer. The internal stress which arises during electrodeposition of the test metal caused the mercury in the capillary to either rise or fall. Mills employed a spherical bulb 11.5 mm in diameter and a second thermometer of the same form and dimensions as the apparatus itself to allow for temperature effect. Calculation of the alteration of the degree of compression or distension of the deposit was made on the basis of the difference in the readings of these two thermometers, the internal stress being proportional to the pressure. Calibration was carried out by placing the thermometers in an apparatus where elevated pressures could be established and the capillary division evaluated in atmospheres. The deposits laid down on Mills' bulb were several millimeters in depth and this must certainly have affected the accuracy of the results. Mills' first data on the internal stresses in electrolytic deposits are presented in Table 14.

The figures of the table are at variance with more recent data, since they indicate that there are very high internal stresses in deposits of silver and gold and comparatively low stresses in deposits of metals of the iron group. Nevertheless, Mills was correct in reporting that electrolytic deposits of copper, iron, nickel, and silver are compressed, whereas deposits of zinc and cadmium are distended.

TABLE 14. Values of the Internal Stress in Certain Metals

Metal	Observed effect	Internal, stress, atmos.
Cd	Tension	2.3
Zn	"	6.2
Ni	Compression	19.2
Fe	"	18.2
Ag	"	66.4
Au	"	90.7

Boutty [2] further developed the Mills method and derived an equation relating the pressure, p, on the glass bulb and the time, consumed in forming the metallic deposit, τ:

$$p = \frac{y + \frac{3u}{5}\xi\tau}{t + \frac{8}{3} \cdot \frac{\xi}{y + \frac{5}{3}\xi} \cdot \frac{y\gamma}{\beta} r^2}. \qquad (16)$$

Here, u is the diminution of the volume of the deposited layer on a cylindrical bulb, y is the diminution of unit volume of the cylinder under unit pressure, ξ is the coefficient of compressibility of the metal, γ is the density of the metal r is the radius of the cylinder, and β is a constant. The data of Boutty yield a β value of $1.226 \cdot 10^{-6}$ for electrolytically deposited copper.

THE METHOD OF CATHODE BENDING

Stony [3] was the first to evaluate the internal stresses in electrolytic deposits from measurements on cathode bending. The internal stress developed in the deposit produces a stress of opposite sign in the base and causes the latter to twist and alter its dimensions. The internal stresses in the base and its electrolytic plate are shown schematically in Fig. 29. Here, the stresses are assumed to be distributed over the section ABCD as indicated, remaining

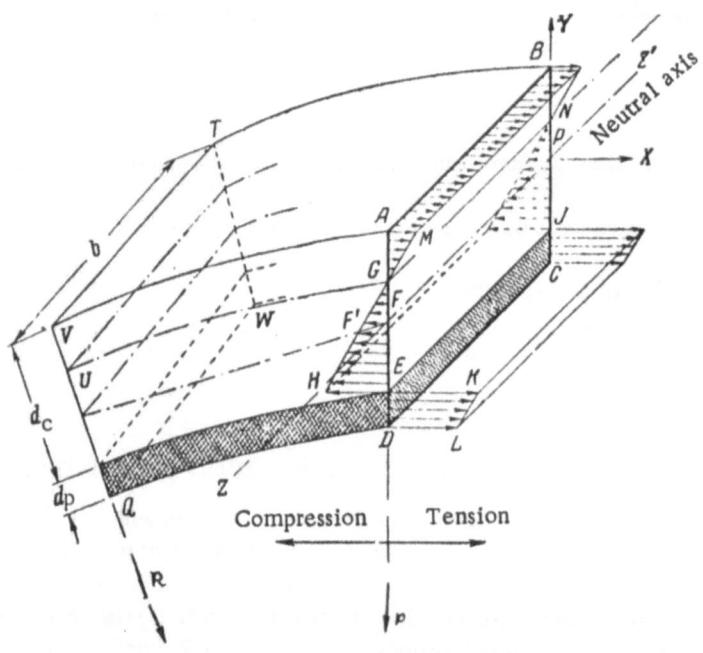

Fig. 29. Schematic representation of the stresses in the base and the electrolytic deposit (Brenner, Senderoff).

constant along the directions AB and CD. The magnitude and direction of the stress in the deposit are indicated vectors drawn from DL to EK. Vectors directed to the left of AD indicate a state of compression in the metal, whereas vectors directed to the right of this same line indicate a state of tension.

It is to be noted that tensions in the deposit give rise to oppositely directed forces in the layer directly beneath the deposit. These forces tend to reduce the volume. Forces similarly directed to those in the deposit arise on the other side of the base. Within the base metal, there will, therefore, be a neutral surface, NGU, whose position is fixed by the ratio of depth of the deposit to the thickness of base and by the force ratio.

Two relations must be fulfilled when the forces in base and deposit are in equilibrium. First, the tension and compression must compensate, so that

$$\int p \cdot dA = 0,$$

dA being an element of the cross sectional area. Second, the bending moment of the specimen around any axis must be equal to zero, so that

$$\int p \cdot y \cdot dA = 0. \tag{18}$$

A graphical representation of these relations can be built up in the following manner. The resultant force will be equal to:

$$\text{(AB} \cdot \text{sum of forces on ED)} + \text{(AB} \cdot \text{sum of forces on GA)}$$
$$-\text{(AB} \cdot \text{ sum of forces on GE)} = 0. \tag{19}$$

The sum of moments will also be equal to zero at equilibrium so that:

$$\text{(AB} \cdot \text{sum of moments on DE)}$$
$$= \text{(AB} \cdot \text{sum of moments on GA)}$$
$$+ \text{(AB} \cdot \text{sum of moments on FE)}$$
$$- \text{(AB} \cdot \text{sum of moments on GF)}.$$

The thickness of the cathode will be designated by d_c, and the depth of the deposit by d_p, with $d_p \ll d_c$. Let it be supposed that the cathode is so bent under the action of the internal stress in the deposit that the radius of curvature is R. Finally, let δ designate the distance measured in the x direction from the edge of the cathode to the neutral axis (Fig. 30). The relative distension of a segment located at a distance δ-x from the neutral axis will, obviously, be:

$$\frac{\delta - x}{R}.$$

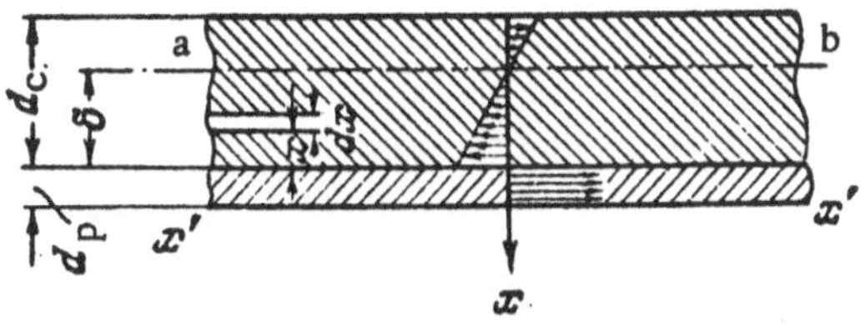

Fig. 30. For evaluation of the internal stress.

The force acting on an element of depth, dx, will be given by:

$$\frac{E}{R}(\delta - x)\,dx, \tag{20}$$

E being the Young modulus. The moment of this force is:

$$\frac{E}{R}(\delta - x)\,x \cdot dx. \tag{21}$$

At equilibrium, the sum of the moments of all the forces in the cathode will be:

$$\int_0^{d_c} \frac{E}{R}(\delta - x)\,x \cdot dx. \tag{22}$$

It follows that $\delta = \frac{2}{3}\,d_c$.

The force acting horizontally on unit cross section as a result of the internal stress at a deposit depth d_p will be:

$$p \cdot d_p = \int_0^{d_c} \frac{E}{R}(\delta - x)\,dx, \tag{23}$$

$$p \cdot d_p = \frac{E}{R}\left(\delta \cdot d_c - \frac{d_c^2}{2}\right). \tag{24}$$

41

Substitution of the expression for δ lead to:

$$p = \frac{1}{6} \frac{E d_c^2}{R d_p} .$$

(25)

Figure 34 shows that:

$$R = \frac{l^2 + 4z^2}{8 \cdot z} ,$$

(26)

where z is the deflection and l is the length of the cathode.

The assumption that z is very small in comparison with l leads to:

$$z = \frac{l^2}{8 \cdot z} .$$

(27)

Replacing R by $l^2/8z$ gives

$$p = \frac{4}{3} \frac{E \cdot d_c^2 \cdot z}{d_p \cdot l^2} .$$

(28)

It has been pointed out above that Stony considered the deposit depth to be small in comparison with the thickness of the base metal in each case.

Stony has given the following equation for deposits of considerable depth:

$$p = \frac{4}{3} \frac{E (d_c^2 + d_p \cdot d_c)z}{d_p l^2} .$$

(29)

Equation (28) has been derived in a similar fashion by Pertsovskii [4].

Mathematical relations between the internal stress and the cathode bending have been derived somewhat differently by Kudryavstev, Saverin, and Ryabchenkov [5] and by Ioffe [6].

The derivation of the above equation relating the internal stress and the bending of the cathode assumed that the deposit depth was small in comparison with the thickness of the base. The error from this source will fall within the limits of experimental accuracy when the deposit depth is no more than 5% of the thickness of the base. There are occasions, however, in which it is necessary to determine the internal stresses in much thicker deposit and here, error of measurement will be more than 50% when the deposit depth is 25% of the thickness of the base. The derivation of the equation also assumes the position of the neutral axis to be fixed at $1/3$ of the distance from the outer surface of the base, and the Young modulus to be the same for the base and the deposit. Failure to take these factors into account leads, at times, to erroneous results. Brenner and Senderoff [7] have studied this matter in detail and have concluded that there are three cases which should be distinguished on the basis of those elastic properties of the base which fix the magnitude of the internal stress in the deposit.

The stresses arising in an elastic base as the result of deposition will deform the deposit and, themselves, be reduced. The true internal stress will be distorted to an extent which will depend on the elasticity of the base.

The first case is that in which the internal stress is determined with a base which has been held rigidly so as to preclude deformation. After deposition of the plate, the base is freed from the restraining forces and allowed to bend until equilibrium is established. The internal stress is then calculated from the curvature.

The internal stress calculated for this case will be close to the true value (Fig. 33). Brenner and Senderoff determine the compressibility of the base through:

$$p \left(\frac{d_p}{d_c + d_p} \right) ,$$

(30)

and the difference between the measured and the true value of the internal stress through:

$$p \left(\frac{d_c}{d_c + d_p} \right) .$$

(31)

42

It is clear that the difference between these two values will increase with the depth of the deposit. Without going into a detailed mathematical derivation, it can be said that the internal stress should be calculated from the equation:

$$p = \frac{E(d_c + d_p)^3}{6 \cdot d_c d_p R} .$$ (32)

The second case is that of deposition on a base which is kept from bending but allowed to contract. The situation here is different from that of the first case, since the internal stress in the base will increase with the depth of

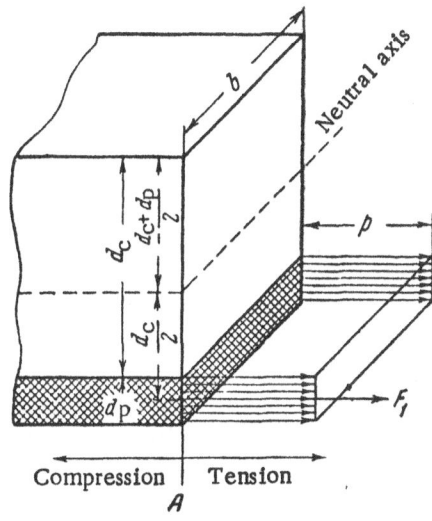

the deposit. The internal stress is determined from a measurement of the radius of curvature of the sheet after removal of the restraining forces. The calculations are complicated by the fact that the tension in one layer tends to compress the next underlying layer as well as the base. For this reason the tension will be greatest in the last layer of the deposit (Fig. 32). In this case, the internal stress will be calculated through the equation:

$$p = \frac{E(d_c + d_p)^3}{3 \cdot R \cdot d_p (2d_c + d_p)} .$$ (33)

A comparison of these two methods shows that the error cannot exceed 10% in the second case.

The third case is that in which deformation of the base entails alteration in the curvature. This is the situation most frequently met in practice. The corresponding equation has the simple form:

$$p = \frac{E \cdot d_c (d_c + d_p)}{6 \cdot R d_p} .$$ (34)

Fig. 31. Schematic representation of the force distribution in deposit and base, Case I (Brenner, Senderoff).

This is Stony's equation with the radius of curvature expressed in terms of the camber $R = l^2 / 8z$.

Figure 33 gives a schematic representation of the force distribution in deposit and base for this third case.

There are times when account must be taken of the fact that the values of Young's modulus for deposit and base are not identical. The difference between these values is considerable in most instances (factor of 10) and becomes especially marked in the case of thick deposits.

Allowance for this difference leads to the following equation for the first case:

$$p = \frac{E_c (d_c + R_1 \cdot d_p)^3}{6 \cdot d_c d_p R} ,$$ (35)

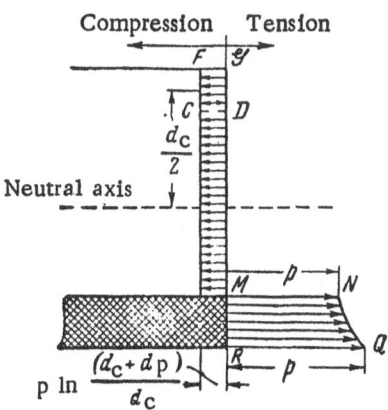

with $R_1 = \dfrac{E_p}{E_c}$.

In the more common case, allowance for this difference replaces Eq. (32) by:

$$p = \frac{E_c d_c (d_c + R_1 d_p)}{6 \cdot R \cdot d_p} .$$ (36)

Fig. 32. Schematic representation of the force distribution in deposit and base, Case II (Brenner, Senderoff)

This last equation generally leads to completely satisfactory results.

It is quite necessary to consider the temperature in determining the internal stress. The difference in coefficients of linear expansion of the deposit and base is another factor which affects the curvature and the magnitude of the internal stress and it, too, must be allowed for when deposition is carried out at elevated temperatures. This calls for the introduction of correction terms which carry Eqs. (32) and (33) into:

$$p = \frac{E \cdot (d_c + d_p)^3}{6 \cdot d_c d_p R} - E \cdot \Delta T \, (A_c - A_p), \tag{37}$$

$$p = \frac{E \cdot d_c \, (d_c + d_p)}{6 \cdot d_p R} - E \Delta T \left(A_c - A_p \left(\frac{d_c}{d_c + d_p} \right)^2 \right). \tag{38}$$

Here, ΔT is the temperature difference and A_c and A_p are the coefficients of linear expansion of cathode and deposit, respectively.

It is of interest to compare the results of calculations based on the Stony Equation with values obtained from the above relations.

Assuming the base and the deposit to have identical moduli of elasticity, the Stony Equation will give results which are in error by 14, 25, 42, and 70% when the d_p/d_c ratio is 0.05, 0.01, 0.2, and 0.5, respectively.

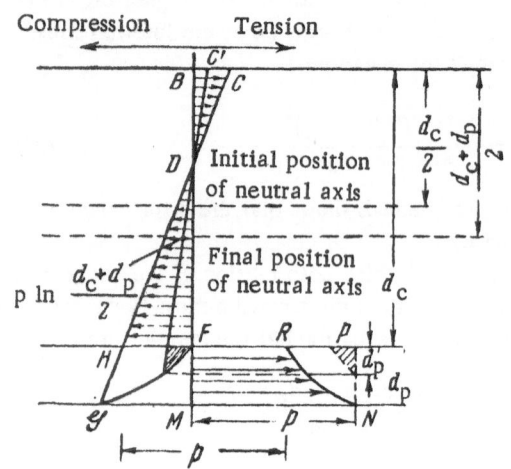

Fig. 33. Schematic representation of the force distribution in deposit and base, Case III (Brenner, Senderoff).

If Eq. (33) is used in place of the Stony Equation, the percentage errors for these same ratios of depth of deposit to base will be 5, 9, 17, and 33%. Thus, the choice of equation to be used in evaluating the internal stress will depend on the method of measurement if high accuracy is desired. In treating thicker deposits, it is also necessary to take account of differences in the linear coefficients of expansion and Young's modulus for base and deposit. Finally, it will be necessary to allow for Young's modulus of the masking film which is used in studying the internal stress by this method.

Gevorkyan [8] has proposed that allowance should be made for the modulus of elasticity of the masking film when calculating the internal stress by either formula, especially when the cathode is rather thin. It is true that the presence of such a film can distort the results, markedly, in some cases. The modulus of elasticity of this film is not significant when the film depth is small in comparison with the thickness of the cathode.

The experimental data are best handled through an equation involving the camber (flexion) rather than the radius of curvature. Figure 34 makes it clear that in terms of the camber $R = \frac{l^2 + 4z^2}{8z}$, whereas, in terms of the displacement of the end of the cathode, z', $R = \frac{l^2 + z'^2}{2z'}$. The quantity z^2 is very small and can be neglected to give $R = \frac{l^2}{8z}$, in the first case and $R = \frac{l^2}{2z'}$, in the second. The δ coefficient of Eq. (28) will be divided by 4, if the displacement of the end of the cathode is used as a measure of deformation instead of the camber. These calculations assume the arc and its chord to be of the same length. The error from this source is generally of the order of 2% and clearly increases with the deflection of the cathode.

The above equations for calculating the internal stress make no allowance for the fact that the deposition of metal takes place on the active sectors of the cathode surface, the entire surface of the base being gradually covered as the individual crystals grow. It is clear that the internal stress does not increase continuously with increasing depth as predicted by the equation. Determinations of internal stresses are carried out without considering this aspect of electrodeposition. The effect of this factor will vary with the conditions of electrolysis.

A determination of the internal stress by the bending method is carried out with a thin metallic sheet cathode, several centimeters in length. This is fixed firmly at one end and is free to move at the other. The anode is a sheet of similar dimensions which is set parallel to the cathode and at a fixed distance from the latter. That side of the cathode which faces the anode is covered with a thin layer of some masking substance such as varnish so that the deposition of metal will occur only on the other side. The internal stress which arises in the deposit will cause the cathode sheet to bend as deposition proceeds, the magnitude and direction of the bending being determined by the magnitude and sign of the stress.

Various pieces of apparatus based on these principles have been described in the literature. These can be separated into two groups, the one containing devices which measure the internal stress after electrolysis is completed, and the other containing devices which permit the stress to be evaluated while electrolysis is in progress.

Attention will now be turned to several devices of the first type which differ principally in the method followed in measuring the cathode bending.

The Macnaughtan Apparatus: The cathode used in this apparatus has the form of two thin sheets which are operated in parallel [9]. Metal is deposited on these sheets to a depth of 1.3 mm. Following this, the sheets are clamped together at the upper end with their nonworking faces in contact; the distance between the lower ends is measured. The results obtained in this manner are poorly reproducible. For this reason the sheets and clamp are heated in boiling water for one hour and the distance then measured again. It is reported that the authors could obtain reproducible results only in this way. The internal stress is determined in millimeters deflection of the cathode from the difference in the distance between the lower ends of the cathodes before and after boiling. The poor reproducibility is probably due to a fact that will be discussed later, namely, that the internal stress and the bending of the cathode both change when the current is cut off for any considerable time. The approach to equilibrium is accelerated by heating in water.

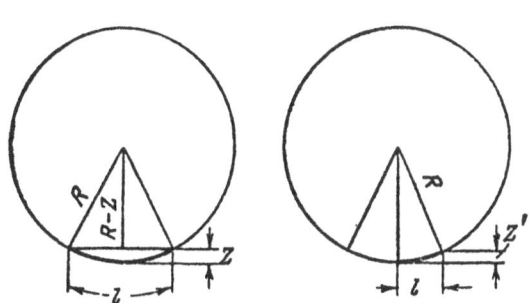

Fig. 34. Diagram for the calculation of internal stress.

The Apparatus of Glikman, Fedot'ev, and Chernova: The vertical optimeter of Glikman, Fedot'ev,

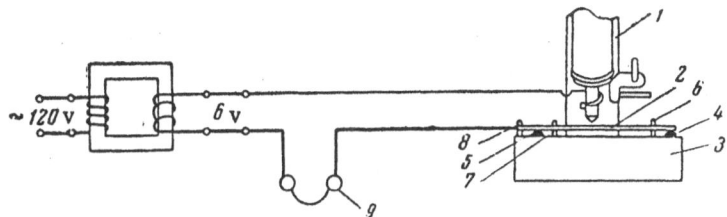

Fig. 35. Schematic representation of an apparatus for measuring cathode bending (Glikman, Fedot'ev, Chernova).

and Chernova [10] is a most refined apparatus for measuring cathode deflection after electrolysis is completed. This device is shown in schematic form in Fig. 35. A measurement of the deflection is carried out by placing the cathode sheet 2 on the two supports 4 and 5 of the special table 3 where it is held fixed by the three stops 6, 7, and 8. The sheet is laid on the supports with the plate side downward and the deflection then measured from the other side by the gauge 1. The telephone receivers 9 are used to record the instant of contact. The normal operation of the entire apparatus is checked against standards in order to avoid effects from temperature variations, such check being made prior to each determination of the curvature of a sheet.

The bending of the sheet is evaluated from the difference in the readings of the optimeter before and after plating. The accuracy of determination of the deflection is 2-3%.

The TsNIITMASh:* An apparatus similar to that of Glikman, Fedot'ev, and Chernova has been developed in the TsNIITMASh [11].

This apparatus consists of a massive metal base, a stand which is rigidly attached to this base and carries a micrometer screw, a contact measuring ring which is set on an insulating panel in the center of the base, and a signal lamp. The set-up is shown schematically in Fig. 36. The micrometer screw is 25 mm long with a pitch of 0.5 mm and carries 0.01 mm graduations. The measuring surface of this screw has spherical form so as to assume that contact with the test sheet is made at a point lying on the screw axis. The null position of the screw is such that the highest point on the measuring sphere is in the plane of the measuring ring. A signal lamp is used to mark out the instant of contact of the screw with the test sheet, the customary divisions of the micrometer screw being too large for 0.1-0.2 mm specimens.

A measurement is carried out with this apparatus by affixing a control disk to the measuring ring and then placing the specimen on this disk (with the concave side downward, if it should happen to be bent). The thickness

---

* The Central Scientific-Research Institute for Machine Construction Technology.

of the specimen is calculated from the reading on the micrometer screw, a, through the equation h = a-10.0 mm, the apparatus being initially adjusted so that the screw reading is 10.

The Marsh Apparatus: The Marsh apparatus for the study of internal stresses is a modification of the apparatus of Philips and Glifton [13] and of Soderberg and Graham [14], based on the principle of measuring the radius of curvature relative to the plane of the sheet. It consists principally of a bakelite adapter and two readily removal supplementary plates which are used to prevent bending of the flexible elastic cathode during electrolysis and, at the same time, improve the current distribution over the cathode surface.

Electrical contact is made through a screw which is in contact with the back side of the cathode. The set screws and electrical leads are insulated with varnish.

The cathode is freed from the retaining plates at the end of the electrolysis and bends, the curvature being fixed by the magnitude of the resultant stress. The radius of curvature is measured with another device which is a modified form of the Phillips assembly using a one-volt lamp instead of the microscrew to mark out the instant at which contact is made with the sheet. It is claimed that the accuracy of measurement is 10%.

The devices of the second group which measure the cathode bending during electrolysis differ from one another in the methods employed for clamping the flexible cathode and determining its bending.

The Stony Method: The simplest variant of the cathode bending method is that which was used by Stony [3] in 1909. Here, metal was deposited on one side of a sheet steel cathode, 102 mm long, 12 mm wide, and 0.32 mm thick, the upper end of this cathode being rigidly clamped above the solution while the lower end was free to move. The cathode deflected under the action of the internal stress and this deflection was measured with a micrometer. The depth of the plating was obtained from the increase in the cathode weight, and the internal stress calculated from the magnitude of the bending.

This method of determining the cathode bending during electrolysis was also used by Kolschütter [15]. Here measurement of the bending of a platinum sheet cathode 6.5 cm long, 2 cm wide, and 0.1 mm thick was made with the aid of a glass needle which was attached to lower end of the cathode and served as a pointer, attachment being made at such an angle that the free end of the needle would project above the level of the electrolyte. This needle indicated the deflection of the cathode, and from this the internal stress could be evaluated.

The Method of Barklie and Davies: Barklie and Davies have studied internal stresses by two methods which they have described only in general terms [16]. In the first of these, the electrolytic cell was placed in a projector and the deflection of the cathode determined from the displacement of its projection on a screen. The distance between the end of the cathode and its projection on the screen was fixed exactly so that it was possible to estimate the sensitivity of the apparatus. Barklie and Davies were able to increase their sensitivity 50-fold by using optical devices. Calculations were carried out on the basis of the difference between the deflection of the cathode after deposition and the deflection after dissolving the deposit in acid.

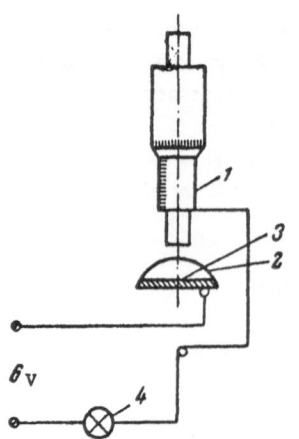

Fig. 36. Schematic representation of an apparatus for measuring camber (TsNIITMASh): 1) Micrometer screw; 2) test sheet; 3) measuring ring; 4) signal lamp.

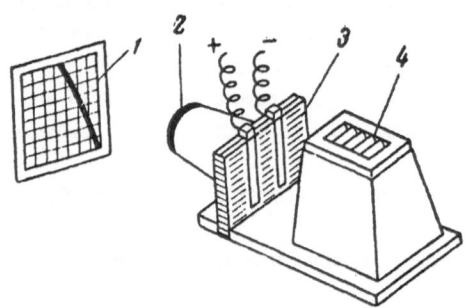

Fig. 37. Schematic representation of an apparatus for studying internal stress (Shluger): 1) Screen; 2) focusing lense; 3) cell; 4) projection lamp.

The second of these methods measured the deflection of the cathode with the aid of a microscope and micrometer scale calibrated in 2.5 mm divisions. The current distribution over the cathode surface was improved by carrying out deposition in a glass cell with plane, parallel sides. The current efficiency for metal deposition was determined simultaneously with the internal stress. The authors point out that this second method is more simple but less accurate than the first.

The Apparatus of Shluger: Shluger has also used projection on a screen to determine the bending of the cathode [17]. His apparatus is shown in schematic form in Fig. 39; it consisted of a thin cell of a colorless transparent plastic, 150 mm high, 250 mm long, and 10 mm wide, containing a sheet steel cathode which was clamped at the upper end. The anode was a thin sheet, 80-100 mm long, which was set parallel to the cathode. A special screening plate of transparent plastic was sealed into the cell with a view to eliminating nonuniformities in the distribution of metal on the cathode surface. Uniformity in distribution was also promoted by having only a small clearance of approximately 1 mm between edge of the cathode and the cell walls.

The cell with its screens was set between the projection lamp and the focusing lense of a diascope and an image of the cathode, magnified 10-fold, projected onto the screen, where the deflection was either measured visually or photographed.

The Method of Marie and Thon: Marie and Thon [18] have also used the cathode bending method, observing the displacement of the free lower end of the cathode through a microscope.

The Apparatus of Pertsovskii: The apparatus proposed by Pertsovskii [4] consists of an electrolytic cell with attachments for clamping the electrode and measuring the deflection; it differs from the devices described above in having the cathode fixed at the lower end rather than the upper.

The current distribution over the cathode surface is improved by inserting the cathode in an oblong housing of transparent plastic which is open above and on the side facing the anode. The clearance between the housing and the cathode is approximately 1 mm so that the electrode can move freely without touching the housing.

Electrical contact is made at the lower end of the cathode where the latter is clamped and insulated by tightly fitting textolite blocks. A thin pointer is attached to the upper end of the electrode and is essentially a continuation of the latter. The weight of this pointer is approximately 0.025 g and its contribution to the bending moment can be neglected for practical purposes.

The apparatus is equipped with a scale graduated directly in millimeters of cathode deflection.

There are still many other devices which have been used for measuring internal stresses; these differ principally in the means employed for following the displacement of the end of the cathode.

Use of the Universal Rack for Measuring Deflections: A very simple and useful apparatus for observing the cathode deflection has been developed in the Institute of Physical Chemistry of the Academy of Sciences of the USSR. This consists of a glass electrolytic cell with attachments for clamping the electrode and making electrical contact, and a universal rack carrying a microscope. The electrode deflection is followed with the microscope, observations being facilitated by allowing the objective to move freely in all directions.

The Apparatus of the Institute of Physical Chemistry of the Academy of Sciences of the USSR: An apparatus for recording the deflection of the end of the electrode during electrolysis has been constructed in the Institute of Physical Chemistry of the Academy of Sciences of the USSR.

This apparatus makes it possible to automatically record the cathode deflection even if the electrolyte is opaque or the gas evolution is excessively high. This cannot be done with other devices. This apparatus is shown schematically in Fig. 38.

The apparatus consists of a rectangular glass cell, 1, with a plexiglas cover 2, onto which there are mounted the electrode clamps, electrical leads, and a device for measuring the bending of the electrode. The cathode is the thin metallic sheet, 6, whose upper end is held in the cover by the copper clamp, 8.

The anode is the metallic sheet 7, or a wire helix; this is also held in the cover by the clamp 8. Current is supplied to the electrodes through the contacts 9.

The lower end of cathode 6, rests on the glass lever assembly, 3, which is attached to the cover by a thin caprone thread and rotates freely about its axis. The other end of this assembly rests on the sheet which carries the mirror 4; this is attached to support 5 by a caprone thread. The internal stress arising during deposition of metal on one side of the cathode causes the latter to bend in one direction or the other relative to the anode, the amount of bending being determined by the conditions of electrolysis. This results in a corresponding

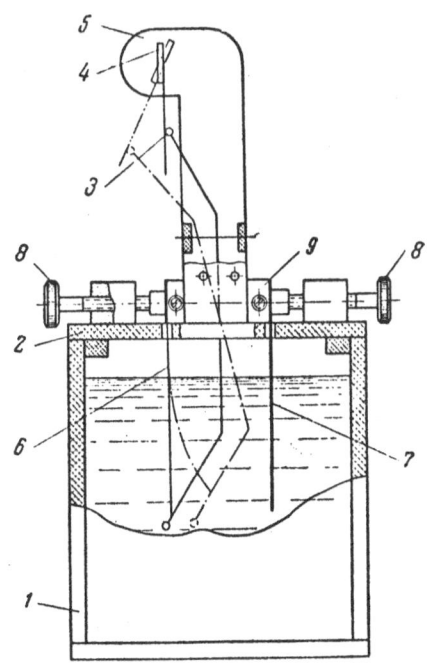

Fig. 38. Apparatus for determining internal stresses (Institute of Physical Chemistry; Tsareva, Vagramyan, Demin).

movement of the lever and mirror. The character and magnitude of the internal stress can be deduced from the movement of a beam of light which is reflected from mirror 4 onto a scale or a film.

The Method of Hoar and Arrowsmith: Hoar and Arrowsmith have proposed a somewhat different apparatus for the measurement of internal stress. This apparatus is shown in schematic form in Fig. 39. The cathode is a metallic bar, 35.5 × 2.5 × 0.32 cm, the lower end of which is clamped in a special holder. The upper end of this bar carries a small mirror which is designed to reflect a ray of light directed upon it and thereby fix the original position of the cathode (null position). The upper portion of the cathode is equipped with two solenoids though which current can be passed so as to hold the cathode in this null position. One of these solenoids is used when the internal stress places the deposit in compression and the other, when the stress gives rise to tension.

Each selenoid is calibrated so that the force required to hold the bar in its null position can be obtained from the selenoid current.

It is unfortunate that reasons beyond our control make it impossible to include a discussion of the Kushner method for determining internal stresses in this section (Proceedings Am. Electroplaters Soc. (1954) p. 188).

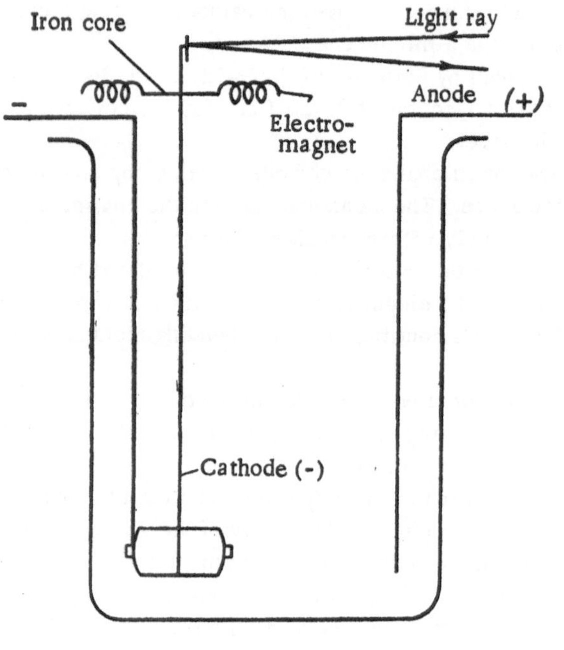

Fig. 39. Schematic representation of an apparatus for measuring internal stresses (Hoar, Arrowsmith).

THE SPIRAL CONTRACTOMETER OF BRENNER AND SENDEROFF

Brenner and Senderoff [7] have suggested that the internal stresses customarily measured are in error because of the contaminants entering the electrolyte from the masking substance which covers one side of the electrode. They have also suggested that the electrode does not usually bend along a circular arc but along a more complex curve, so that this is another source of difficulty in evaluating the internal stress. For these reasons they have proposed a new device, the so-called spiral contractometer, for determining internal stresses. The principal feature of this apparatus is the cathode. This is prepared from a rectangular metallic sheet which is wound spirally on a rod and clamped to the latter at both ends. Deposition occurs on one side only of this sheet, the other side being screened by the base rod.

The helical cathode will either twist or untwist depending on the nature of the stress set up in the deposit, and this stress is evaluated from a measurement of the angular displacement. A schematic drawing of the apparatus is given in Fig. 40.

The authors have proposed two variants of this apparatus, one with a horizontal, and the other with a vertical dial.

Any alteration in the diameter of the helical cathode is mechanically communicated to a pointer which moves over a calibrated scale.

The magnitude of the internal stress can be determined from the diameter, pitch, and height of the helix, the thickness of the cathode sheet, and the depth of the deposit. The equation used in these calculations is a slightly modified form of the Stony relation, namely:

$$p = \frac{E d_c^2}{6 d_p} \cdot \Delta\left(\frac{1}{R}\right). \tag{39}$$

Here p is the internal stress, E is Young's modulus for the base metal, d_c is the thickness of the cathode sheet, d_p is the depth of the electrolytic deposit, and, $\Delta(1/R)$ is the alteration in curvature resulting from the deposition.

$$\Delta\left(\frac{1}{R}\right) = \frac{D \cdot \Pi}{\pi \lambda h}. \tag{40}$$

The D of this last equation is the angular deflection of the helix, Π is its pitch, λ is its external diameter, and h is its height.

Equations (39) and (40) give:

$$p = \frac{E d_c^2 \Pi}{6 \pi \lambda h} \cdot \frac{D}{d_p} \,. \tag{41}$$

The quantities E, d_c, λ, h, and Π are all constant, d_p is obtained from the difference in the weight of the cathode before and after electrolysis, and D is determined from the movement of the pointer.

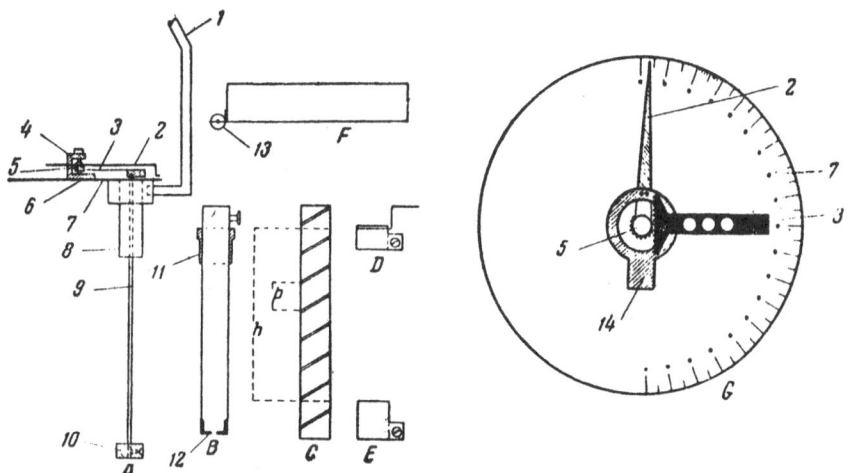

Fig. 40. Schematic representation of the apparatus of Brenner and Senderoff: A) Basic assembly; B) guard tube; C) helix; D) top clamp with solder wire lead; E) bottom clamp with lead; F) plastic cap with movable calibration assembly; G) top view of dial face; 1) hanger; 2) pointer; 3) lever arm; 4) bearing housing; 5) gear on pointer; 6) gear segment on lever; 7) dial face; 8) collar; 9) torque rod; 10) insulating base with set screw; 11) collar; 12) bearing; 13) removable pulley for calibration; 14) balancing counterweight on pointer.

It has been pointed out above that the derivation of Eq. (41) assumes the depth of the deposit to be much less than the thickness of the base metal. Cases in which the depth of the deposit cannot be neglected require the replacement of this equation by:

$$p_T = \frac{E (d_c^2 + d_p \, d_c)}{6 d_p} \, \Delta \left(\frac{1}{R} \right). \tag{42}$$

p and p_T are related through:

$$p_T = p \left(1 + \frac{d_p}{d_c} \right). \tag{43}$$

If allowance is to be made for Young's modulus, E_p, of the deposit, Eq. (43) must be replaced by:

$$p_T = p \left(1 + \frac{E_p}{E} \cdot \frac{d_p}{d_c} \right). \tag{44}$$

It is clear that

$$p_T = p.$$

when the d_p / d_c ratio approaches zero.

Brenner and Senderoff claim that one of the great advantages of using the spiral contractometer for measuring internal stresses is that experimental errors arising from the variability in the physical properties of the electrode metal can be eliminated by determining the deflection of the spiral at the beginning of each experiment. This determination is based on the equation:

$$M = k\bar{D}, \tag{45}$$

in which M is the moment of inertia of the spiral, and D is the angular deflection of the pointer. Once the value of the constant k is known, the internal stress can be calculated from the equation:

$$p = \frac{2k}{\Pi d_c} \cdot \frac{\bar{D}}{d_p} . \tag{46}$$

Young's modulus can also be determined with the spiral contractometer:

$$E^* = \frac{12\pi\lambda hk}{\Pi^2 d_c^3} \cdot g, \tag{47}$$

where

$$g = \frac{\bar{D}}{D} .$$

Brenner and Senderoff have shown that their method gives trustworthy results which are easily reproducible when the current is distributed uniformly over the spiral. This last condition is reached by correct alignment of the anode.

A similar apparatus has been developed by Frumer and Chistov [21].

The Spiral Contractometer of Poperek: The spiral contractometer of Poperek [22] is a modified form of the apparatus of Brenner and Senderoff and differs from the latter in using an optical system to replace the gear and pointer.

The author claims that this makes possible a 15- fold increase of sensitivity over the best values obtained with the apparatus of Brenner and Senderoff.

THE STRAIN GAUGE METHOD

The Peperek and Frusin method uses a disk cathode and a resistance strain gauge; it could be placed in a third group.

Such gauges are widely used in various automatic, industrial devices such as those employed for measuring strain in stressed machine parts.

Poperek and Frusin [23] have employed these gauges for determining the internal stresses in electrolytic deposits. Their apparatus is shown in schematic form in Fig. 41a. The working gauge 1 is affixed to the inactive side of cathode 2 which is clamped to the metallic body 3 of the apparatus by ring 4. The anode 7 is set parallel to the cathode and held to the walls of the apparatus at fixed distance by vinyl plastic screws 6. The gauge is screened from below by the cover 5 which is attached to the body of the apparatus.

The body of the apparatus is varnished inside and then filled with the electrolyte.

The electrical circuit is shown schematically in Fig. 41b. Current from the storage battery 1 passes into the electrolytic cell 3 through rheostat 2. The working gauge 4 is attached to the cathode. A compensating gauge 5 is inserted in one arm of a bridge, fed from the battery 6 and the working gauge in the other arm. The difference of potential between the terminals 7 of the measuring diagonal is recorded with a sensitive instrument (10^{-8}-10^{-9} amp/ mm) or is fed into an amplifier.

By using these gauges it is possible to measure the strain which arises on the inactive side of the disk cathode because of bending under the action of the internal stress in the deposit.

Poperek and Frusin calculate this internal stress p, through the equation:

$$p_0 = p_D \cdot \frac{9E_c \, d_p}{8\mu_p(1+\mu_p)(1+\mu_c)} \left(\frac{1-\mu_c^2}{144E_c \, d_c} + \frac{A}{9E_p \, d_p} \right), \tag{48}$$

---

* Equation (47) is obtained rrom (41) through:

$$p = \frac{2k}{\Pi d_c} \cdot \frac{\bar{D}}{d_p} .$$

50

p_0 being the initial value of the stress, p_D the stress obtained from a tensometer,* E the modulus of elasticity of the deposit, d_c the thickness of the cathode, μ_p is the Poisson coefficient of the deposit, μ_c the Poisson coefficient of the cathode, d_p the depth of the deposit, and $A \approx 3$.

The working gauge and the compensation gauge are held under strictly controlled conditions in order to avoid thermal and other effects.

A similar apparatus has also been developed by Potapov and Sanzharovskii [24].

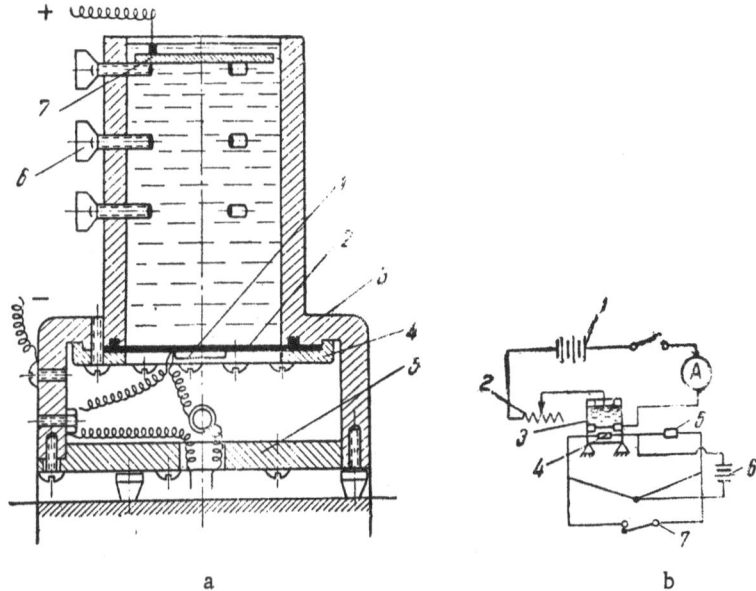

Fig. 41. Apparatus of Poperek and Frusin for determining internal stresses by the strain gauge method. a) Diagram of the apparatus; b) main schematic.

THE X-RAY METHOD OF DETERMINING INTERNAL STRESS

The methods discussed so far lead to the determination of the "total" stress in the electrolytic deposit.

The x-ray method of determining internal stress is different in so far as it permits separate determination of the so-called stresses of the first and second types [25].

Stresses of the First Type: It is known that each stress invariably brings about some deformation of the body in question. Stresses of the first type act in the deformation of an assemblage of crystal grains distributed over a finite volume and functioning as an isotropic body. Such stresses are characteristic of elastic deformation. Here, each grain is subject to the same deformation and the interplanar spacing is the same throughout the entire grain, differing from the normal value by an amount Δd_M. The simplest case is that of linear deformation where the relative change $\Delta d_M / d_M$ is proportional to the stress p:

$$\frac{\Delta d_M}{d_M} = \mu \cdot \frac{p}{E} . \tag{49}$$

Here d_M being the interplanar distance, μ is the Poisson coefficient, and E is the modulus of elasticity.

The stress, p, can be obtained from the known values of μ and E and an x-ray determination of the $\Delta d_M / d_M$ ratio.

The accuracy of this method increases with the angle of incidence, θ, of the ray and is dependent on such other factors as:

a) The error in measuring the angle

---

* The tensometer is used to estimate the force of tension or compression from the deformation caused by the forces.

b) the error in determining the distance between the specimen and the photographic plate;

c) the temperature variation during x-raying.

Stresses of the Second Type: The interplanar distance in a polycrystal under plastic deformation is not the same for all crystals but varies from one grain to another, the relation between the deformations of the individual grains being quite complex. It can be considered that such body is subjected to a irregular deformation and one then speaks of microstress or stress of the second type. The x-ray diagrams obtained in cases of internal stress of the second type are characterized by the diffuseness of the interference lines. The difficulties involved in measuring stress of the second type are discussed in detail in the book of Kitaigorodskii [25].

It should be remarked that only a very limited amount of work has been done on the x-ray determination of the internal stresses in electrolytic deposits. The paper of Palatnik, Gepshtein, and Lukov [26] on the quantitative determination of the internal stresses in electrolytic zinc deposits and their relation to the conditions of electrolysis was one of the first in this field. These authors have noted that the relative experimental error with respect to the value of the internal stress is quite high. This experimental error is reduced considerably in the more highly perfected, x-ray method recently proposed by Mamontov and Petrov [27].

LITERATURE

1. E. J. Mills, Proc. Roy. Soc. 26, 507 (1877).
2. E. Boutty, Compt. rend. 88, 714 (1879); 92, 868 (1881).
3. G. G. Stony, Proc. Roy. Soc. 82, 172 (1909).
4. M. L. Pertsovskii, Porous Chrome Plating [in Russian] (Mashgiz, Moscow-Sverdlovsk, 1949).
5. I. V. Kudryavtsev, M. M. Saverin, and A. V. Ryabchehkov, Methods of Surface Hardening of Machine Parts [in Russian] (Mashgiz, Moscow, 1949).
6. V. S. Ioffe, Uspekhi Khim. 8, 50 (1944).
7. A. Brenner and S. Senderoff, J. Res. Nat. Bur. Standards 42, 89 (1949).
8. V. M. Gevorkyan, Zhur. Fiz. Khim. 33, 1318 (1959).
9. D. J. Macnaughtan and A. W. Hothersall, Trans. Faraday Soc. 24, 387 (1928); 31, 1168 (1935).
10. L. A. Glikman, N. P. Fedot'ev, and A. P. Chernova, Zavodskaya Lab. 17, 1126 (1951).
11. A. V. Rykova, Studies on the Corrosion of Metals under Stress [in Russian] (Mashgiz, Moscow, 1953).
12. V. J. Marchese, J. Electrochem. Soc. 99, 39 (1952).
13. W. H. Phillips and F. L. Flifton, Proc. Amer. Electroplaters Soc. 34, 97 (1947).
14. G. Soderberg and A. K. Graham, Proc. Amer. Electroplaters Soc. 34, 74 (1947).
15. V. Kolschüter, Z. Elektrochem. 24, 300 (1918).
16. R. H. D. Barklie and J. J. Davies, Engineer 150, 670 (1930).
17. M. A. Shluger, Collection, The Theory and Practice of Electrolytic Chrome Plating [in Russian] (Acad. Sci. USSR Press, Moscow, 1957) p. 147.
18. C. Marie and N. Thon, J. Chem. Phys. 29, 11 (1932).
19. Yu. S. Tsareva, V. G. Solokhina, N. T. Kudryavtsev, and A. T. Vagramyan, Zhur. Fiz. Khim. 29, 166 (1955).
20. T. P. Hoar and D. Z. Arrowsmith, J. Electroplaters and Metal Finish, 10, 141 (1957).
21. L. A. Frumer and N. M. Chistov, Zavodskaya Lab. 24, 244 (1958).
22. M. Ya. Popereka, Zavodskaya Lab. 23, 6 (1957).
23. M. Ya. Poperek and K. S. Frusin, Vestn. Akad. Nauk Kaz. SSR 2, 84 (1958).
24. G. K. Potapov and A. T. Sanzharovskii, Zhur. Fiz. Khim. 32, 1416 (1958).
25. A. I. Kitaigorodskii, X-rays and X-ray Methods [in Russian] (Gostekhizdat, Moscow, 1948).
26. L. S. Palatnik, E. Gepshtein, and I. B. Lukov, Zhur. Tekh. Fiz. 10, 1756 (1940).
27. E. A. Mamontov and Yu. N. Petrov, Uchenye Zapiski Leningrad. Gosudar Pedgog. Inst. im. A. I. Gertsen, Ser. Fiz. i Matem. 141 (1958).

Chapter IX

THE EFFECT OF VARIOUS FACTORS ON THE INTERNAL STRESS

THE EFFECT OF THE NATURE OF THE DEPOSITED METAL

It is very difficult to systematize the metals in terms of deposit stress, since this value is quite sensitive to a number of factors. A comparison of this kind can be made only if it is possible to carry out each deposition under exactly the same conditions and this cannot be done in most cases. Since the magnitude of the internal stress is highly dependent on the salt which is employed, the same type of salt must, naturally, be used throughout in comparing internal stresses in various metals. It is unfortunate that many investigators have failed to take this into account. The internal stress is strongly dependent not only on the anion, but also on the other components of the solution, the conditions of electrolysis, and still other factors. Thus, it is quite difficult to use the literature data as a basis for arranging the metals in terms of internal stress. Table 15 gives the data of various authors on certain metals.

The table includes cases in which there is more than a twofold variation in the values given for a single metal. Thus, the internal stress in copper is 500 kg/cm^2 according to Ioffe and 150 kg/cm^2 according to Fedot'ev and Khonikevich. Similar variations are found in the data on other metals.

It is, of course, possible to arrange the metals in terms of the internal stress (kg/cm^2) but the above remarks make it clear that the ordering is only provisional:

$$Pd \quad > \quad Cr \quad > \quad Ni \quad > \quad Mn \quad > \quad Cu \quad > \quad Zn$$

Pd	Cr	Ni	Mn	Cu	Zn
7000	3700—6000	2800—2900	300—460	150—500	(—20—260)

Most metals are put into tension by the internal stress, but there are cases in which compression results. Thus, electrodeposits of chromium, nickel, cobalt, iron, palladium, and copper are generally in tension, while deposits of zinc, lead, and cadmium, are usually in compression.

THE EFFECT OF THE NATURE AND STATE OF THE BASE

The problem of the effect of the base on the internal stress in the deposit is quite important, since electrodeposition is almost always made on to a different kind of metal. The matching of base to deposit is very significant, both for the adhesion of deposit to base and for the corrosion protection that the deposit affords.

The effect of the base on the internal stress becomes especially significant in the case of thin platings. The existence of a relation between the structure of the electrolytic deposit and the nature and structure of the base has been repeatedly reported in the literature [11]. The internal stress is closely related to the structure of the deposit and changes when this structure is altered. Finch, Wilman, and Yang [12] have expressed the opinion that the effect of the base varies with the depth of the deposit. These authors divide the deposit into three arbitrarily defined layers:

1) A layer directly adjacent to the base whose structure is completely determined by the base. This layer is, at most, a fraction of a micron in depth.

2) A second, transition, layer whose depth does not exceed 5 μ.

3) A third layer whose structure is fixed by the composition of the electrolyte and the conditions of electrolysis.

The structure of the base will be reproduced in the first layer of the electrolytic deposit, if the lattice parameters of base and deposited metal are not radically different. Unless this conditions is fulfilled, deposition proceeds through the formation of three-dimensional nuclei, just as on an inert electrode.

The theory of orientation crystallization acquires great interest in connection with these ideas.

The Dankov [13] theory of orientational and dimensional correspondence yields the following expression for the maximum value $(\Delta\alpha_1)_{max}$ of the difference between the lattice parameters of base and deposit:

$$(\Delta\alpha_1)_{max} = \left(\frac{2a_1\sigma}{c_{11} + c_{12}}\right)^{1/2}.$$

Here α_1 is the lattice constant, σ is the surface tension, and c_{11} and c_{12} are the coefficients of elasticity of the two-dimensional nuclei.

Bliznakov [14] has extended this work to allow for the variation of $(\Delta\alpha_1)_{max}$ with the nature of the base, the degree of supersaturation, and the value of the minimum critical temperature required for orientation crystallization and has shown that $(\Delta\alpha_1)_{max}$ should be calculated from the equation:

$$(\Delta\alpha_1)_{max} = \left(\frac{2a_1\sigma}{c_{11} + c_{12}}\right)^{1/2}\left(1 + \frac{\kappa_b - \kappa - A_{def}}{kTG}\right)^2. \tag{51}$$

Here G is the degree of supersaturation, κ is the work required for separating two neighboring particles in the crystal lattice, κ_b is the work required for separating a particle from the base, and A_{def} is the work of deformation per particle.

This same author has also proven that a thin metallic deposit laid down on a clean electrode will alter its structure as the structure of the base changes. This fact must also bear on the internal stress. It is unfortunate that the effect of the base structure on the structure and internal stress in the metallic deposit has not yet been studied under purely laboratory conditions.

The electrodes used in applied electrolysis are deformed by the mechanical treatment to which they are subjected and the effect of the base on the internal stress in the deposit is thereby masked to a considerable degree. The same can also be said of action of those films of oxides and other foreign substances which partially cover the surface and strongly affect the structure and internal stress of the deposit.

It must be noted that there is only a limited body of experimental data on the effect of the base on internal stress. The one detailed study in this field is that of Marie and Thon [15] in which it was shown that the internal stress in nickel is strongly dependent on the cathode metal:

TABLE 14.

	Al	Ta	Cu	Co	Ni	Sn	Ag	Au	Fe	Pd
Lattice parameter, A	1.43	—	1.27	1.26	1.26	1.4	1.44	1.44	1.26	1.37
Internal stress, relative values	0.7	1.0	2.0	2.7	3.7	5.0	5.5	5.7	6.0	13.7

Comparison shows that there is a certain parallelism between lattice parameters and internal stresses in metals.

The breakdown of this parallelism in iron, aluminum, and palladium is probably to be explained by the fact that the surface state distorts the effect of the base metal, the surface being Pd—H in the case of palladium [16] and a metallic oxide in the case of aluminum or iron.

It is quite likely that a different ordering would be found from the deposition of other metals.

A study by Rykova [1] has shown that the internal stress in deposits of considerable depth (about 20 μ) is no longer affected by the nature of the base, but is strongly dependent on the surface state of the latter. The effect of mechanical surface treatment on the internal stress in nickel deposits obtained from the electrolyte of Table 15 is illustrated by the following figures:

Base	Treatment of surface	Camber, mm
Copper	Polished	3.0
"	"	2.5
"	"	2.5
"	Abraded	4.0

Base	Treatment of surface	Camber, mm
Copper	Abraded	4.5
"	"	4.0
Brass	Polished	3.0
"	"	2.5
"	"	2.0
"	Abraded	4.0
"	"	5.0
"	"	5.0
Steel	Polished	2.5
"	"	2.5
"	"	2.5
"	Abraded	4.5
"	"	4.0
"	"	4.0

These data show that there is higher internal stress in a nickel deposit laid down on an abraded surface than in a similar deposit laid down on a polished surface, both deposits being obtained from the above electrolyte at a current density of 1 amp/dm^2 and a temperature of 20°C.

THE VARIATION IN INTERNAL STRESS AFTER CUTTING OFF THE CURRENT

The foregoing has brought out the fact that a variation in the deposit depth must lead to a change in the internal stress. This is true regardless of whether or not deposition has been made on to a base of the same metal, since the deposit undergoes a certain degree of structural change as a result of the inevitable passivation of the electrode surface. These conclusions have been confirmed experimentally by Brenner and Senderoff [17].

Figure 42 (Curve 1) makes it clear that the value of the internal stress is quite high at the beginning of electrolysis and then sinks to a stationary value as the depth of the deposit increases. Thus the internal stress in nickel takes on a practically constant value at a depth of 0.013 mm. Brenner and Senderoff explain the diminution of the internal stres with increasing depth in terms of the increase in the dimensions of the crystals in the electrolytic deposit. These authors have studied the deposition of metals under various conditions and have found a parallelism between internal stress and deposit structure.

Kushner [18] has also thoroughly investigated the variation of internal stress with depth in the electrodeposition of nickel under various conditions and has proposed that these quantities be related through the following empirical equation:

$$p_a = A + B + e^{-k \cdot d}p. \tag{52}$$

Here p_a is the mean value of the internal stress at a depth d_p, A is the stationary value of the internal stress under the given experimental conditions, B is a constant characterizing the base metal, and k is another constant.

This equation predicts that the internal stress will alter exponentially with the deposit depth.

The laws applying to structural change in the deposit vary with the current strength and this factor must be taken into account in establishing the relation between depth and internal stress.

THE VARIATION OF THE INTERNAL STRESS WITH THE DEPOSIT DEPTH

A number of authors have made the interesting observations that the internal stress continues to increase even after the current has been cut off [19]. The observed change in stress is always in the positive direction regardless of the type of stress generated in the deposit during passage of the current. This increase continues as long as 30 min in certain cases, depending on the nature of the metal and the conditions of deposition. Figure 43 shows the variation of the internal stress in nickel deposits during passage of the current and after the current has been cut off. The internal stress is found to change abruptly at the instant of cutting off the current, in some instances. Kushner [20] has estimated that the stress can vary from its original value by 2-17% after cutting off the current, the relative change being the greater the lower the absolute magnitude. Kushner explains this relation by supposing that the diffusion layer gives rise to a reverse current after the circuit has been opened, so that there is a dissolution of the deposit on certain sectors of the electrode and further deposition on others. The result is that the surface

TABLE 15. The Internal Stresses in Various Metals

Metal	Electrolyte composition	Temperature of electrolyte, °C	Current density, amp/dm²	Depth of deposit, micron	Internal stress, kg/cm²	Author
Chromium	CrO_3—250 g/liter H_2SO_4—2.5 g/liter	65	80.0	30	+3700	Rykova [17]
Chromium	CrO_3—250 g/liter H_2SO_4—2.33 g/liter	65	—	3	+4300	Pertsovskii [2]
Nickel	$NiSO_4 \cdot H_2O$—250 g/liter H_3BO_3—30 g/liter NaF — 8 g/liter KCl — 3 g/liter	20	—	—	+1810	Barklie and Davies [3]
Nickel	Ni_{met}— 40 g/liter or $NiSO_4 \cdot 7H_2O \approx 190$ g/liter Na_2SO_4—18 g/liter H_3BO_3— 20 g/liter pH — 4.5–5.7 NaCl — 50 g/liter	20	1	—	+2800–2900	Rotinyan and Kozich [4]
Nickel	$NiSO_4 \cdot 7H_2O$—250 g/liter $MgSO_4 \cdot 7H_2O$— 30 g/liter H_3NO_3 — 30 g/liter KCl — 3 g/liter NaF — 3 g/liter	20	4	30	+2800	Rykova [17]
Copper	$CuSO_4 \cdot 5H_2O$—250 g/liter H_2SO_4 — 5 g/liter	18	1.0	75	+230	Fedot'ev and Pozin [5]

TABLE 15 (Cont'd)

Metal	Electrolyte composition	Temperature of electrolyte, °C	Current density, amp/dm²	Depth of deposit, micron	Internal stresses, kg/cm²	Author
Copper	$CuSO_4 \cdot 5H_2O$ — 250 g/liter H_2SO_4 — 5 g/liter	18-20	2.0	–	+500	Ioffe [6]; Fedot'ev and Khonikevich [7]
Zinc	$ZnSO_4 \cdot 7H_2O$ — 143 g/liter CH_3COONa — 0.25 g/liter gum arabic — 1 g/liter	18	–	–	+150	Barklie and Davies [3]
Zinc	$ZnSO_4 \cdot 7H_2O$ — 450 g/liter $Al_2(SO_4)3 \cdot 18H_2O$ — 30 g/liter	20-23	1-10	–	–20-120	Gorbunova and Popova [8]
Palladium	20-5-1.2 g/liter metallic Pd; Na_2HPO_4 — 100 g/liter $(NH_4)_2HPO_4$ — 20 g/liter NH_4Cl — 25 g/liter aqueous ammonia silution added to pH — 9,0	18-20	–	–	+7000	Ostroumov [9]
Manganese	$MnSO_4 \cdot 4H_2O$ — 175 g/liter $(NH_4)_2SO_4$ — 100 g/liter SO_2 — 0.3 g/liter pH — 3.0	20	25.0	–	+300	Potapov and Sanzharovskii [10]
Manganese	The same; pH 4.2	20	25.0	–	+300	Potapov and Sanzharovskii [10]

57

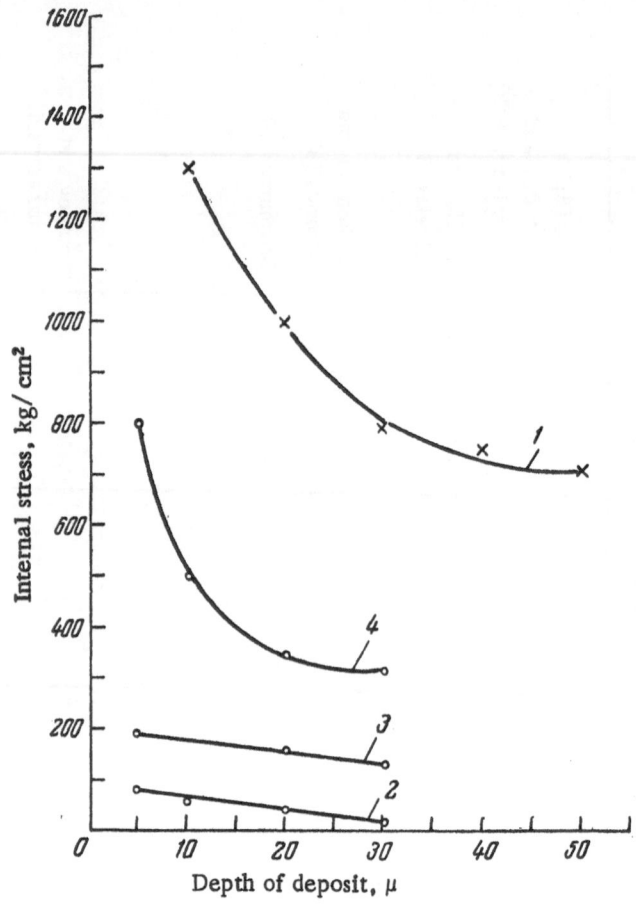

Fig. 42. The relation between depth and the internal stress in deposits of zinc and nickel: 1) Nickel (Brenner and Senderoff); 2) zinc, I_c = 1.0 amp amp/dm^2 (Gorbunova, Popova); 3) zinc, I_c = 10.0 amp/dm^2 (Gorbunova, Popova); 4) zinc, I_c = 3.0 amp/dm^2 (Gorbunova, Popova). Deposition of zinc from an electrolyte containing: $ZnSO_4 \cdot 7H_2O$, 215 g/liter; potassium alum, 30 g/liter; thiourea, 1.25 g/liter.

layer acquires a finely crystalline structure and the internal stress increases. It seems to us that this effect might also be due to the fact that the temperature of the deposit falls after the current is cut off, and the volume of the deposit is reduced thereby. There is the further possibility of a reduction in volume as the result of the escape of hydrogen, especially from the surface layers.

THE EFFECT OF CURRENT DENSITY

The density of the current used in preparing the deposit is of great significance in fixing the internal stress and this fact is commonly drawn on to obtain plates with minimal stress. The relation between internal stress and current density varies from metal to metal.

Chromium

It has been pointed out above that the internal stresses in electrolytic chromium deposits are so high that cracking occurs and the experimental results obtained from stress studies are, thereby, distorted. This fact makes for difficulty in establishing the absolute value of the internal stress and the exact effect of current density. It must be noted, also, that the relation between internal stress and current density varies with the temperature. Figure 44 gives internal stress, current density curves which have been obtained by Shluger [21] in an electrolyte with the composition: CrO_3, 150-170 g/liter; CrO_3/H_2SO_4, 100-120. The figure shows that the internal stress increases with the current density. The data of Rykova [1] on chromium deposits obtained from a solution containing 250 g/liter CrO_3 and 2.5 g/liter H_2SO_4 indicate that the relation between these quantities is a complex one (Fig. 47). The data of Pertsovskii [2] also prove that these factors are not related simply.

The work of Glikman, Fedot'ev, and Chernova [22] has proven, however, that the current density exerts only a minor effect on the internal stress.

It is clear from what has been said that the current density, internal stress data obtained by different authors working under somewhat different conditions are contradictory and the intercorrelation of these factors must, there-

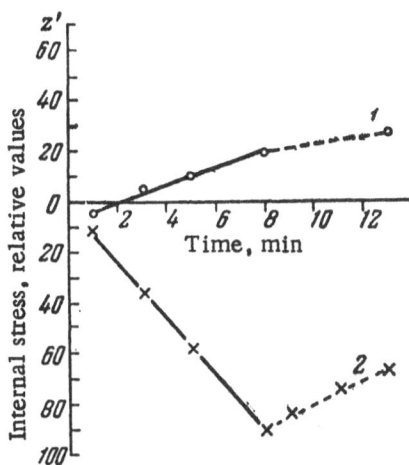

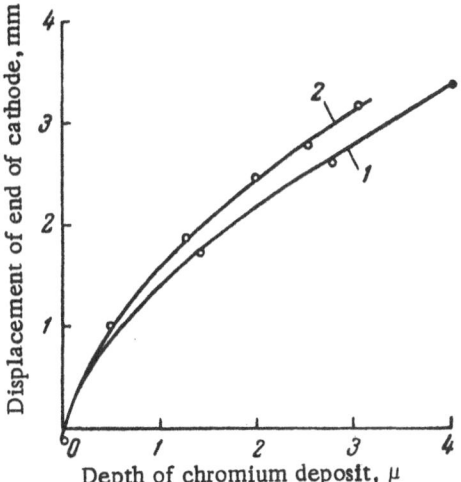

Fig. 43. Time variation of the internal stress in electrolytic nickel deposits during current flow and after cutting off the current (Vagramyan, Petrova): $I_c = 3.0$ amp/dm^2 and $t = 18-20°$: 1) Variation of internal stress in the presence of 0.2 g/liter of thiourea; 2) variation of internal stress in the presence of 0.6 g/liter of thiourea. Deflection in millimeters equal to z' · 8.16 · 10^{-2}.

Fig. 44. The relation between current density and the internal stress in chromium deposits (Shluger). Composition of electrolyte: $CrO_3 = 150-170$ g/liter; $[CrO_3]/[H_2SO_4] = 100-120$; $t = 60°$; 1) $I_c = 30$ amp/dm^2; 2) $I_c = 60$ amp/dm^2.

fore, be quite complex. It is very likely that these contradictions are due, in part, to the fact that the deposit cracks once its depth has reached a certain minimum value and the results are distorted thereby.

It is, in fact, true that the internal stress in thin deposits increases with the current density, although cracking causes the stress to fall off at high density even here.

Nickel

In electrodeposits of nickel the current density, internal stress relation varies with the depth, just as does the corresponding relation for chromium. Curves showing the variation of internal stress with depth in nickel deposits are given in Fig. 46. From these it is seen that the internal stress is a linear function of the depth in deposits obtained at low densities. Departures from linearity are observed in thicker deposits obtained at higher current densities, and these appear earlier the higher the current. Figure 46 has been used for constructing graphs showing the relation between current density and internal stress for deposits of 1 μ depth (Fig. 47, 1) and 2.5 μ depth (Fig. 47, 2). A comparison of these curves indicates that the internal stress rises gradually with the current density in the thin deposit, while the curve for the deposit of 2.5 μ depth passes through a maximum. It is possible that these results are due to the structural alteration which occurs in passing from shiny to dull deposits and to the elimination of the effect of the base on the structure of the deposited metal.

The effect of current density on internal stress has been studied by a number of other authors. Thus Samartsev and Lyzlov [19] have investigated the relation between internal stress and current density in solutions containing: $NiSO_4 · 7H_2O$, 200 g/liter, H_3BO_3, 25 g/liter, and KCl, 5 g/liter at pH 5.7 and have shown that the curve passes through a minimum. This work contains no indication of the deposit depths corresponding to the reported figures.

Rotinyan and Kozich [4] consider the increase in internal stress with current density to be related to that increase in alkalinity of the layer surrounding the electrode which accompanies the deposition of nickel. This alkalization is greater at the higher current densities and is supposed to account for the increase in internal stress. These authors claim that prevention of alkalization in this layer not only eliminates the increase in internal stress with current density but actually reduces the stress somewhat, variation of the current density being then practically without effect.

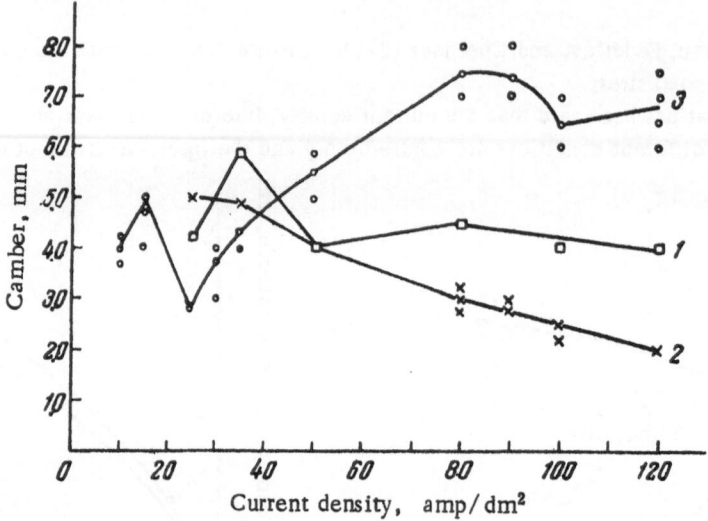

Fig. 45. The relation between current density and cathode camber (Rykova): Temperature: 1) 50°; 2) 55°; 3) 65°. Chromium deposits obtained from an electrolyte containing: CrO_3, 250 g/liter; H_2SO_4, 2.5 g/liter.

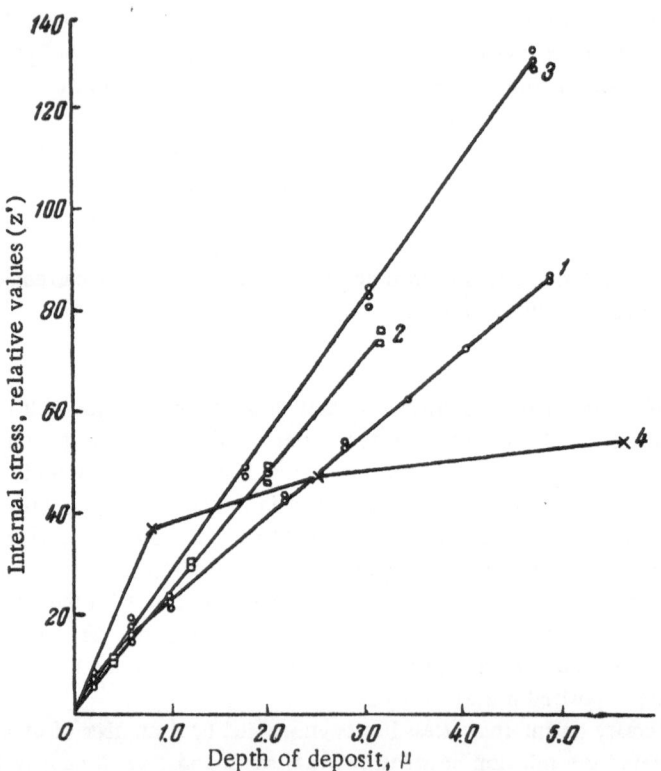

Fig. 46. The relation between internal stress and depth of nickel deposits (Vagramyan, Petrova). Current density: 1) 1 amp/dm²; 2) 2 amp/dm²; 3) 3 amp/dm²; 4) 4 amp/dm². Nickel deposit obtained from an electrolyte containing: $NiSO_4 \cdot 7H_2O$, 266 g/liter, H_3BO_3, 30 g/liter; $Na_2SO_4 \cdot 2H_2O$, 24 g/liter; $MgSO_4 \cdot 10H_2O$, 20 g/liter, pH = 3.78, t = 18 - 20°. Deflection in millimeters equal to z' $\cdot$ 8.16 $\cdot$ 10⁻².

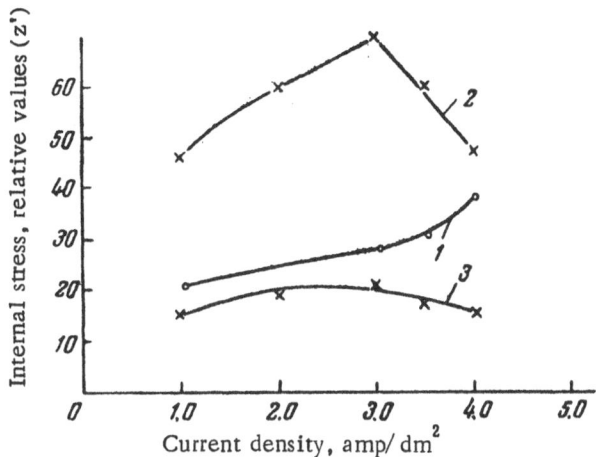

Fig. 47. The relation between current density and internal stress and copper deposits (Vagramyan, Petrova). 1) Depth of nickel, 1 μ; 2) depth of nickel, 2.5 μ; 3) depth of copper, $\approx$ 3 μ. Copper deposits obtained from an electrolyte containing: $CuSO_4 \cdot 5H_2O$, 250 g/liter; H_2SO_4, 50 g/liter; t = 18-20°. Deflection in millimeters equal to z' $\cdot$ 8.16 $\cdot$ 10^{-2}.

The data of Rykova [1] shows an increase in the internal stress with current density (Fig. 48), although the stress is reduced markedly by agitation.

Copper

The curve showing the variation of the internal stress in copper with current passes through a maximum (Fig. 47, 3) just as does the corresponding curve for nickel [23]. The fact that there is a fall-off in internal stress at certain values of the current density is possibly due to the formation of loose, bulky deposits of poor quality at both low and high densities. A good deposit with high internal stress is obtained at a definite current density. The higher the concentration of the salt of the depositing metal, the higher the current density required for producing dense, high-quality deposits. It is observed here that the maximum on the internal stress curve is displaced toward higher current densities. The relation between internal stress and current density has been studied by Fedot'ev and Kruglova for the case of the electrodeposition of copper in the presence of Rochelle salt [24] and it has been shown that the stress increases with current density up to 2.0 amp/dm^2.

Zinc

The studies of Gorbunova and Popova [8] on electrolytic zinc deposits have again shown that the internal stress rises with the current density (Fig. 42, Curves 2, 3, and 4). The curve for the thin deposit has a much greater slope than does the curve for the thick deposit. This difference is explained by the fact that the thin deposit is a compact polycrystalline mass while the thick deposit is built up from larger, poorly adhering, crystals and is therefore friable.

These examples show that the internal stress rises with an increase in the current density. The fact that the internal stress begins to fall off once a certain current density has been reached is to be ascribed to such extraneous factors as cracking of deposits with high internal stress and structural changes in deposits with low internal stress.

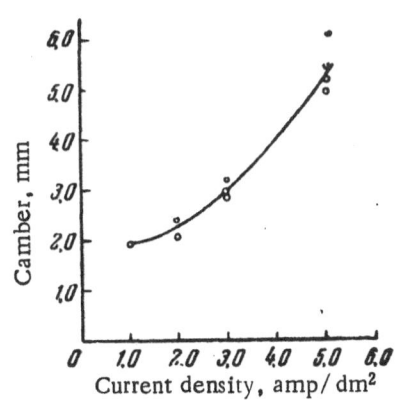

Fig. 48. The effect of current density on cathode camber (Rykova). Nickel deposits obtained from an electrolyte containing: $NiSO_4 \cdot 7H_2O$, 250 g/liter; $MgSO_4 \cdot 10H_2O$, 30 g/liter; H_3BO_3, 30 g/liter; NaF, 3 g/liter; KCl, 3 g/liter. pH = 4.3, t = 45°.

There is a connection between the rise in internal stress with increasing current density and the increase in the overvoltage of the metal with the establishment of conditions under which a nonequilibrated crystal lattice is formed. It follows that those metals which deposit out at high overvoltage should also show high internal stresses, and conversely. This conclusion is confirmed by the experimental data of Table 16 on the overvoltage and internal stress in the deposition of various metals from various salts. The current density intervals were so selected here that a compact homogeneous deposit would be obtained in each case.

TABLE 16. Internal Stress and Overvoltage for Various Metals

Metal	Composition of solution and conditions of electrolysis	Overvoltage, mv	Internal stress, relative values
Zinc	Sulfate electrolyte, $I_c = 2.0$ amp/dm^2	40	-0.42
Copper	Sulfate electrolyte, $I_c = 3.0$ amp/dm^2	50-168	$+(0.9-1.65)$
Nickel	Sulfate electrolyte, pH 2.5; $I_c = 3.0$ amp/dm^2	500-550	$+(4.9-6.5)$*
Cobalt	Sulfate electrolyte, pH 2.5; $I_c = 2.5$ amp/dm^2	255	$+5.1$
Palladium	Ammonium chloride electrolyte, $I_c = 0.5-2.0$ amp/dm^2	800-850	$+(5.75-9.7)$*

*In each of these deposits the internal stress was reduced by the formation of a crack system.

This table indicates that the overvoltage and internal stress diminish in the order: Pd > Ni > Co > Cu > Zn, high overvoltage being associated with high stress. It should be noted that the effect of the nature and state of the base must be eliminated if the existence of such a parallelism is to be confirmed definitely.

THE EFFECT OF TEMPERATURE

The temperature dependence of the internal stress varies from metal to metal and its study is complicated by the fact that temperature changes entail alteration in such factors as the current efficiency for metal deposition and the adsorption of surface-active substances. The existence of various types of temperature effect could be anticipated since occlusion follows various mechanisms and the internal stress is closely related to the amount of material taken up by the deposit.

The lack of concordance in the reported data on the temperature—internal stress relations in metals has arisen from the fact that solutions of various compositions and pH values have been used in this work and observation made at various deposit depths.

Studies on electrodeposits of chromium have shown that one type of temperature—internal stress relation applies to thick deposits and another type to thin deposits [21]. Figure 49 shows the effect of temperature on the internal stress in electrodeposits of chromium 1, 7, and 12 μ deep.

An increase in temperature reduces the internal stress in the 1 μ deposit (Curve 1), possibly by diminishing the amount of occluded hydrogen. The opposite type of relation is observed in the thick deposit (Curve 3).

The fact that the internal stress increases with rising temperature is indication that there is an alteration in the degree of cracking of the deposit in this case.

The cracking of the deposit is more extensive at low temperatures and the diminution in the internal stress more pronounced for this reason. The cracking diminishes as the temperature rises and the relative reduction falls off as a result. Both factors come into play in deposits of intermediate depth and cause the curve to pass through a minimum (Curve 2). It should be noted that Curves 1, 2, and 3 are not comparable since they are based on internal stress values expressed in millimeters deflection and refer to deposits of different depths.

The temperature—internal stress relation is found to be more uniform in electrodeposits of nickel, cobalt, copper, silver, and certain other metals. Stony's work [25] gives an internal stress of 2700-3000 kg/cm^2 for electrolytic nickel deposits obtained at 10-15°, this value falling to 1200-1500 kg/cm^2 when the temperature is raised to 80-90°. The observations of Stager [26] indicate that a more coarsely crystalline deposit with lower internal stress is obtained at elevated temperature. Rykova [1] has also confirmed the fact that rising temperature diminishes the internal stress in electrodeposits of nickel obtained at pH 4.3 from a solution containing: $NiSO_4 \cdot 7H_2O$, 250; $MgSO_4 \cdot 7H_2O$, 30; H_3BO_3, 30; NaF, 3; and, KCl, 3 g/liter (Fig. 50).

Similar results have been obtained by the present authors. The data of Kushner [18] also show the internal stress in nickel deposits to fall off with increasing temperature (Fig. 51). There is indication, however, of a complex relation between temperature and internal stress in deposits of this kind.

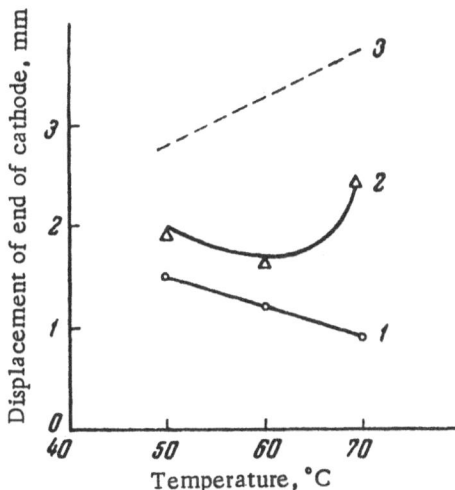

Fig. 49. The relation between temperature and internal stress in chromium deposits depth: 1) 1 μ; 2) 7 μ; 3) more than 12-15 μ (Shluger).

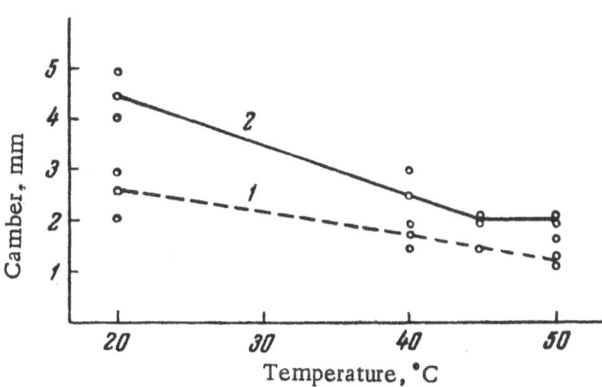

Fig. 50. The relation between temperature and camber (Rykova): 1) I_c = 1.0 amp/dm² with deposition of nickel continued for 1 hr; 2) I_c = 1.0 amp/dm² with deposition of nickel continued for 2 hr.

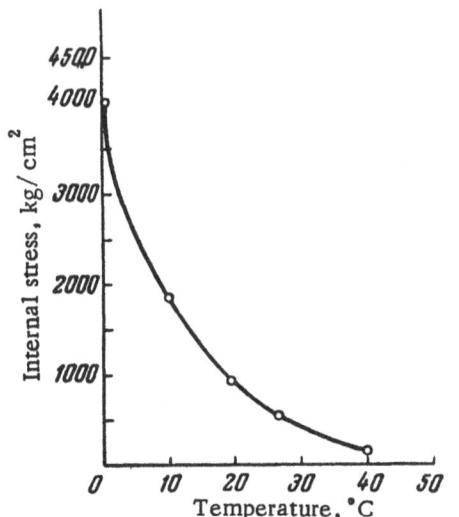

Fig. 51. The relation between temperature and internal stress in nickel deposits (Kushner).

For other metals, an elevation of the deposition temperature usually leads to a diminution of the internal stress. A study by Kolschütter and Jakober [27] has brought out the existence of a parallelism between grain coarseness and internal stress in copper and silver deposits.

THE EFFECT OF THE ACIDITY OF THE ELECTROLYTE

The effect of the pH of the solution on the internal stress has been studied repeatedly, especially in the case of nickel. The data of Samartsev and Lyzlov [19] show that the internal stress in nickel deposits obtained at a current density of 0.1 amp/dm² from a solution containing: $NiSO_4 \cdot 7H_2O$, 200; H_3BO_3, 25; and, KCl, 5 g/liter is essentially constant up to pH 6. Further increase in the pH leads to a marked rise in the stress (Fig. 52).

Our own study of the effect of pH on the internal stress in nickel deposits obtained from an electrolyte containing: $NiSO_4 \cdot 7H_2O$, 166; H_3BO_3, 30; Na_2SO_4, 24; and, $MgSO_4$, 20 g/liter has given somewhat different results, showing a minimum stress at pH 4.73 (Fig. 53).

It is to be noted that these data apply to nickel deposits obtained from solutions containing the nickel salt and still other components. The pH dependence of the internal stress is found to be of a different type in nickel deposits obtained from pure salt solutions, the change of stress with pH being quite pronounced here. It is difficult to extend these studies to cover an extensive pH interval, the range for the preparation of compact homogeneous deposits being quite narrow. This relation can be studied quantitatively in electrodeposits of iron and cobalt, however. Curve 1 of Fig. 54 shows the relation between pH and internal stress in deposits obtained from a 1 N cobalt sulfate solution. It is seen from this curve that the internal stress falls off markedly in the neighborhood of pH 3.0. Similar results have been obtained on iron deposits prepared from 1 N iron sulfate solutions. It is striking that the course of the curve is independent of the nature of the salt, even though the absolute value of the internal stress varies considerably (Curve 2).

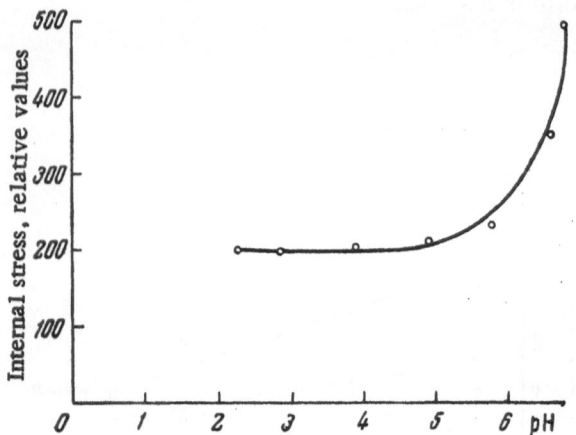

Fig. 52. The relation between the pH of the electrolyte and the internal stress in nickel deposits (Samartsev, Lyzlov). Deposits obtained from a solution containing: $NiSO_4 \cdot 7H_2O$, 200; H_3BO_3, 25; and, KCl, 5 g/liter at a current density of 0.1 amp/dm^2.

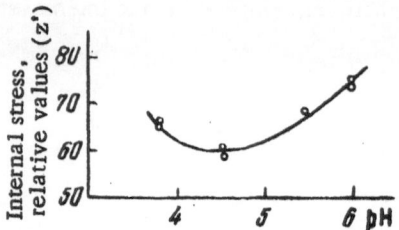

Fig. 53. The relation between the pH of the electrolyte and the internal stress in nickel deposits (Vagramyan, Petrova): I_c = 2.0 amp/dm^2, t = 18-20°. Deflection in millimeters equal to z' $\cdot$ 8.16 $\cdot 10^{-2}$.

A study by Kushner [18] has shown that the internal stress in a nickel deposit depends on the salt from which deposition has been made (Table 17).

The effect of pH on the internal stress in palladium deposited from an ammonium chloride* electrolyte containing 28.8 g/liter of palladium is very peculiar, since the curve showing stress as a function of concentration of free ammonia passes through a minimum (Fig. 55). The figure makes it clear that minimum internal stress in such deposits occurs at a free ammonia concentration of 5.5 g/liter, regardless of the current density.

The data of Fedot'ev and Kruglova [24] show that the stress diminishes with increasing acid concentration in copper deposits obtained at a current density of 1.0 amp/dm^2 from a solution containing $CuSO_4 \cdot 5H_2O$, 200 g/liter and Rochelle salt, 0.2 g/liter (Fig. 56).

It is evident from these remarks that the form of the pH-internal stress relation varies widely with the nature of the metal and the composition of the solution.

THE EFFECT OF FOREIGN SALTS

Study has proven that the addition of foreign salts to the electrolyte will markedly affect the internal stress in the deposit. Table 18 gives the data of Vuilleumier [28] on the electrodeposition of nickel in the presence of foreign salts.

This table brings out the fact that the internal stress in nickel deposits is increased by addition of ferrous sulfate to the electrolyte, increased still more by the addition of sodium chloride, and sharply reduced by the addition of zinc sulfate. An illustration of the relation between internal stress and the nature and concentration of the added salt is found in Table 19 which gives the data of Vuilleumier [28] on the electrodeposition of nickel in the presence of ammonium sulfate and boric acid.

These data make it clear that ammonium salts increase in the stress markedly.

Ioffe [29] has studied the effect of variation of the sodium chloride concentration on the internal stress in electrodeposits of nickel. The data covered by the curves of Fig. 57 indicate that the internal stress rises with increasing

---

* This ammonium chloride electrolyte contained (g/liter):

Catholyte	Anolyte
$Pd(NH_4)Cl_2 - 40$	$(NH_4)_2CO_3 - 10$
$NH_4OH - 30$	$(NH_4)_2SO_4 - 20$
$NH_4Cl - 10$	$NH_4OH - 45$

and was used at I_c = 1.0 amp/dm^2.

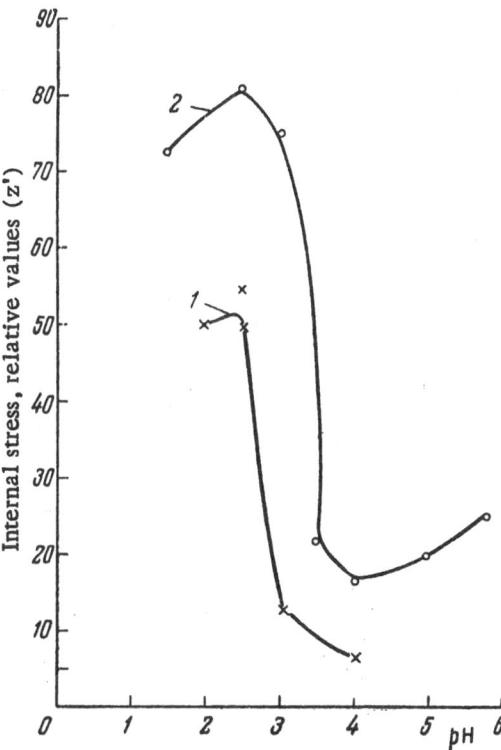

Fig. 54. The relation between the pH of the electrolyte and the internal stress in cobalt deposits obtained from various salts (Vagramyan, Petrova):
1) Cobalt deposit obtained from 1 N $CoSO_4$ at I_c = 2.0 amp/dm^2 and 25°;
2) deposit obtained from 1 N $CoCl_2$ at I_c = 2.0 amp/dm^2 and 25°. Deflection in millimeters equal to z' · 8.16 · 10^{-2}.

TABLE 17. The Relation between the Nature of the Salt and the Internal Stress in Nickel Deposits

Salt	Depth of deposit, μ				
	2.54	5.00	8.00	13.00	51.00
	internal stress, kg/cm^2				
Fluoborate	1100	1415	1283	1228	1187
Bromide	1408	1208	1083	924	780
Sulfate	1932	1883	1725	1656	1596
Chloride	2573	2284	2227	2277	2277

NaCl concentration, reaches a definite limiting value, and then remains constant. The hardness of the nickel deposit rises with the internal stress. The study of Rotinyan and Kozich [4] on the internal stress in nickel prepared in the presence of NaCl has given similar results, the stress varying linearly with concentration over the interval from 5 to 100 g/liter. This effect of NaCl is explained by the reduction which occurs in the deposit grain dimensions as the salt concentration is increased. The authors believe this diminution to arise from the fact that chloride ions adsorb on the more active sectors of the cathode and prevent further crystal growth.

Vozdvizhenskii [30] has shown that nickel deposits prepared in the presence of selenium dioxide are so highly stressed that cracking results in them. It is the opinion of this author that the high stress observed here is due to extensive hydrogenation resulting from catalytic activity in the selenium dioxide.

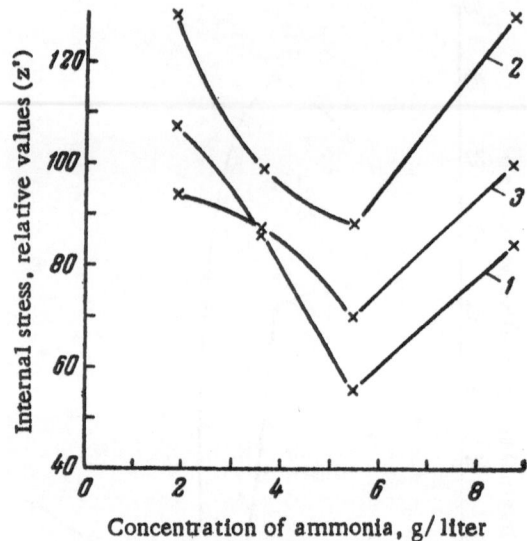

Fig. 55. The relation between the concentration of free ammonia in the electrolyte and the internal stress in palladium deposits (Vagramyan, Petrova). Palladium deposited from an electrolyte containing: $Pd(NH_4)_2Cl_2$, 40 g/liter; NH_4OH, 30 g/liter; NH_4Cl, 10 g/liter. 1) $I_c = 0.25$ amp/dm^2.

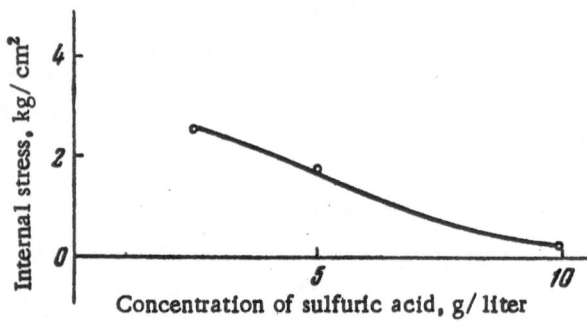

Fig. 56. The relation between the concentration of sulfuric acid and the internal stress in copper deposits (Fedot'ev, Kruglova). Copper deposited from a solution containing: $CuSO_4 \cdot 5H_2O$, 200 g/liter; Rochelle salt, 0.2 g/liter at $I_c = 1.0$ amp/dm^2.

TABLE 18. The Effect of Foreign Salts on the Internal Stress in Nickel Deposits

Duration of electrolysis, min	Internal stress, cathode bending, mm			
	without additives	0.005 N FeSO$_4$	0.005 N ZnSO$_4$	1 N NaCl
0	0	0	1.0	0
5	5.3	5.0	1.0	9.0
10	9.0	10.8	1.9	15.2
15	11.0	16.0	3.0	20.5

TABLE 19. The Effect of the Nature and Concentration of the Salt on the Internal Stress in Nickel Deposits.

Concentration of $NiCl_2$, N	Concentration of added salt, N	pH of solution	Duration of electrolysis, min	Internal stress, relative values (cathode bending, mm)
1.0	0.08 H_3BO_3	3.6	10.0	26.5
1.0	0.08 H_3BO_3	3.6	15.0	91.5
1.0	1.0 NH_4Cl	3.6	10.5	59.0

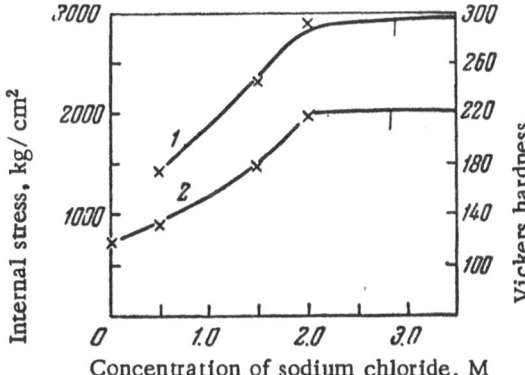

Fig. 57. The relation between the NaCl concentration and the internal stress in nickel: 1) Vickers hardness; 2) internal stress (Ioffe, Stroganov, Grubina).

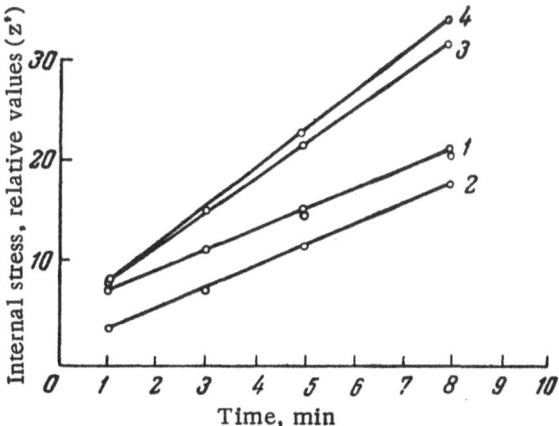

Fig. 58. The time variation of the internal stress in copper deposits in the presence of added foreign salts (Vagramyan, Petrova): 1) In the pure electrolyte; 2) 1.0 g/liter added Ni; 3) 1.0 g/liter added Zn; 4) 1.0 g/liter added Al Deflection in millimeters equal to $z' \cdot 8.16 \cdot 10^{-2}$.

A study of the effect of zinc, aluminum, and nickel sulfates on the internal stress in electrodeposits of copper [31] has shown the stress to be increased by the first two salts and diminished by the third, comparison being made with the stress in a deposit obtained from an additive-free electrolyte (Fig. 58).

These facts indicate that the internal stress can be either increased or reduced by inorganic salts. Similar effects from foreign salts are observed in the deposition of nickel. The change in the internal stress of copper resulting from the presence of zinc, aluminum, and nickel salts is explained by the fact that these metals are occluded in the deposit and distort the crystal lattice; the data of Rosenheim [32] showing the lattice parameter of copper to be increased by the occlusion of aluminum or zinc, and diminished by the occlusion of nickel.

Fedot'ev and Kruglova [24] have also studied the effect of foreign salts on the internal stress in copper deposits and have shown the change caused by Rochelle salt to be quite marked.

The results obtained by Ioffe [6] in a study of the effect of iron salts on the internal stress in nickel are shown in Fig. 59. It is obvious from this figure that such salts increase in the stress, sharply.

THE EFFECT OF SURFACE-ACTIVE SUBSTANCES

It is well known that surface-active substances are widely used for obtaining electrodeposits of high luster and enhanced hardness. The introduction of a surface-active substance into the electrolyte also affects the internal stress; this matter has been treated in numerous papers. The effect of a surface-active substance on the internal stress is dependent on such factors as: the nature, concentration, and purity of the additive; the stability of the additive under current flow and with the current turned off; and the nature and pH of the electrolyte.

The reduction in additive concentration which arises from occlusion in the deposit, reduction at the cathode, or oxidation, at the anode, frequently make for difficulty in interpreting the effect of the additive on the internal stress.

The curves of Fig. 60 show the time variation of the internal stresses, first, in copper deposits prepared in the presence of freshly prepared naphthalenedisulfonic acid and thiourea (Curve 1) and then, in other deposits prepared from an electrolyte which had been stored for a year (Curve 2) [31].

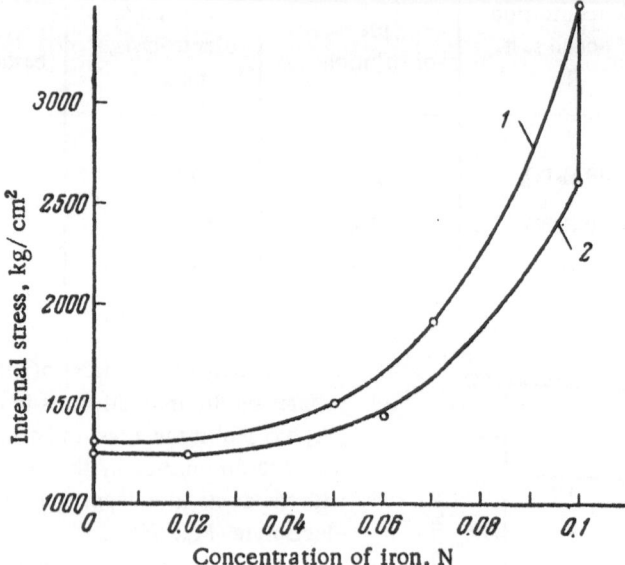

Fig. 59. The effect of iron salts on the internal stress in nickel deposits (Ioffe): 1) 25°; 2) 55°.

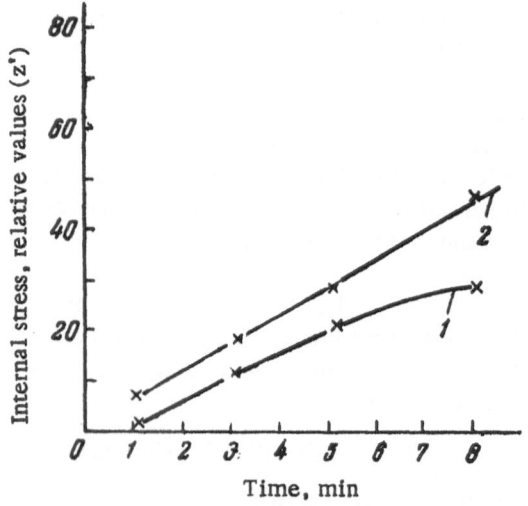

Fig. 60. The effect of additive aging on the internal stress in nickel deposits (Vagramyan, Tsareva): 1) 0.005 g/liter thiourea + 0.5 g/liter naphthalenedisulfonic acid (freshly prepared); 2) 0.005 g/liter thiourea + 0.5 g/liter naphthalenedisulfonic acid (after extended storage). Deflection in millimeters equal to $z' \cdot 8.16 \cdot 10^{-2}$.

The figure makes it clear that the internal stress in copper obtained from a freshly prepared solution containing naphthalenedisulfonic acid and thiourea is much lower than the stress in the deposit obtained from a solution which had been stored for an extended period. It is obvious that this difference reflects the fact that naphthalenedisulfonic acid oxidizes during storage to give oxidation products which are more effective than the pure additive itself.

A similar effect is noted when using additives of varying degree of purity. A relatively small amount of impurity can be more effective than the additive itself and may even bring about a change in the sign of the stress. It is, for this reason, quite important that the composition and purity of the additive be determined. The curves of Fig. 61 show how the use of additives of various degrees of purity effects the time variation of the internal stress in copper deposits. It is seen from this figure that the stress with technical grade anthracene sulfate is twice as great as the stress with the pure compound.

The internal stress can increase, diminish, or even change its sign, depending on the nature of the additive. It should be noted, however, that the effect of a surface-active substance also varies with the conditions of electrolysis.

Thus the internal stress in copper electrolytically deposited in the presence of thiourea can be changed from tension to compression by altering the concentration of the additive and the current density [31]. The relation between thiourea concentration and internal stress in copper deposited at room temperature and $I_c = 3.0$ amp/dm² is shown in Fig. 62. This figure indicates that there is a gradually diminishing tensile stress at thiourea concentrations less than 0.25-0.26 g/liter and a gradually increasing compressive stress at higher concentrations.

Similar results have been obtained by Khonikevich and Fedot'ev [7] in a study of the effect of the concentration of thiourea on the internal stress in copper deposits, by Fedot'ev and Kruglova [5] in a study of the deposition of copper in the presence of Rochelle salt, and by Samartsev and Lyzlov [19] in a study of the deposition of nickel in the presence of sulfonated naphthalene. It should also be noted that a change in the current density at fixed concentration of additive can bring about a marked alteration in the internal stress [31]. The curves of Fig. 63 show that electrolytes with 0.6 g/liter of added thiourea give rise to compressive stresses at cathodic current densities of 1.0-4.5 amp/dm^2 and tensile stresses at current densities of 4.5-6.5 amp/dm^2, the most pronounced change in stress occurring at densities in the range 3-5 amp/dm^2.

The additives can, nonetheless, be separated into three groups, namely:

1) additives which increase the internal stress;

2) additives which have no affect on the internal stress;

3) additives which lower the internal stress.

Ioffe [6] has studied the effect of various organic compounds on the internal stress in nickel deposits obtained from an electrolyte containing: NiSO$_4$, 155; KCl, 15; and, H$_3$BO$_3$, 20 g/liter at pH 5-5.5 and has found that such substances as tannin, analine, nitroanaline, nitrophenol, quinone, pyridine, and strychnine give rise to very high internal stresses, and even cracking, at concentrations as low as 10 mg/liter. Other organic compounds such as benzoic and salicylic acids, and glucose have no effect on the internal stress, even when they are present in the electrolyte at high concentration.

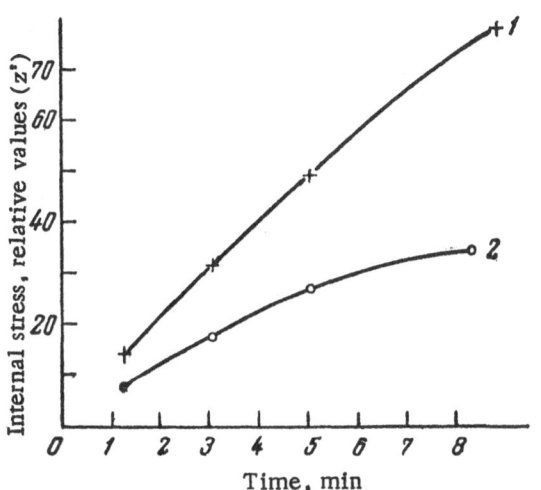

Fig. 61. The time variation of the internal stress in copper deposits corresponding to various degrees of purity of the additive (Vagramyan, Tsareva): 1) 0.1 g/liter technical grade sulfonated anthracene; 2) 0.1 g/liter pure sulfonated anthracene; standard electrolyte; temperature, 18-20°; $I_c = 3.0$ amp/dm^2; deflection in millimeters equal to z' $\cdot$ 8.16 $\cdot$ 10^{-2}.

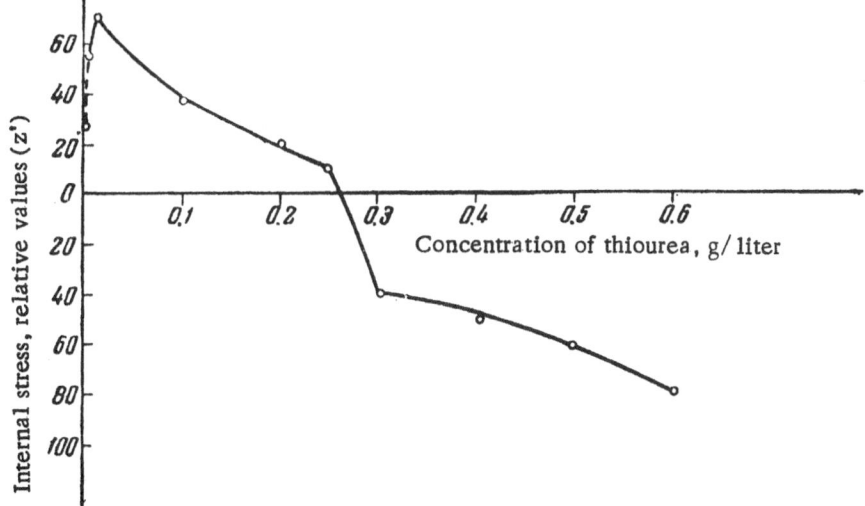

Fig. 62. The relation between the concentration of thiourea and the internal stress in copper deposits (Vagramyan, Tsareva): $I_c = 2.0$ amp/dm^2; t = 18-20°. Deflection in millimeters equal to z' $\cdot$ 8.16 $\cdot$ 10^{-2}.

Ise [33] has studied the effect on the internal stress of certain surface substances which are widely used to improve the quality of industrial nickel deposits. The additives tested were: p-toluenesulfonamide (CH$_3$C$_6$H$_4$SO$_2$NH$_2$, concentration, 171 g/liter) which reduces the internal stress markedly, methylquinolinum chloride [C$_9$H$_6$N(CH$_2$)Cl, concentration, 0.0079 g/liter] which promotes luster, sodium benzylnaphthalenesulfonate ($C_6H_5CH_2C_{10}H_6SO_3Na$, concentration, 0.96 g/liter) which also promotes luster, and sodium lauryl sulfate ($C_{12}H_{25}SO_4Na$, concentration, 1.44 g/liter) which eliminates pitting.* Electrolysis was carried out without agitation in an electrolyte containing $NiSO_4$, 300; $NiCl_2$, 45; and, H_3BO_3, 30 g/liter at pH 5.8 and 40°. The author reports that the internal stress is affected very little by the change in current density resulting from the presence of these additives. The additive which eliminates pitting also has little effect on the internal stress (Fig. 64). The addition of p-toluenesulfonamide leads to a sharp reduction in the internal stress. Methylquinolinum chloride increases the luster markedly, but has very little effect on the internal stress and the author, therefore, concludes that there is no connection between these two characteristics.

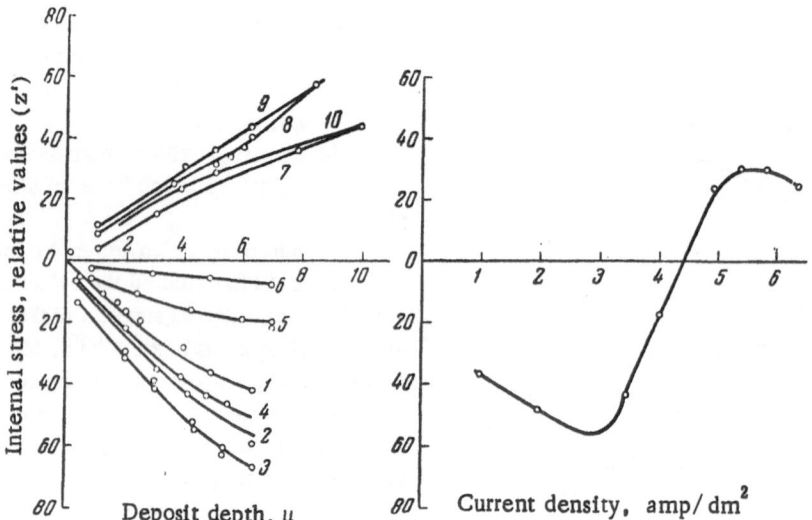

Fig. 63. The relation between the current density and the internal stress in copper for the case of an electrolyte containing 0.6 g/liter added thiourea (Vagramyan, Tsareva): a— 1) I_C = 1.0 amp/dm²; 2) I_C = 2.0 amp/dm²; 3) I_C = 3.0 amp/dm²; 4) I_C = 3.5 amp/dm²; 5) I_C = 4.0 amp/dm²; 6) I_C = 4.5 amp/dm²; 7) I_C = 5.0 amp/dm²; 8) I_C = 5.5 amp/dm²; 9) I_C = 6.0 amp/dm²; 10) I_C = 6.5 amp/dm²; b) curve for a 5 μ deposit constructed from the curves of a.

Both the luster and internal stress are increased markedly by the addition of sodium benzylnaphthalenesulfonate and become closely dependent on the current density. The internal stress diminishes in mixtures of sodium benzolnaphthalenesulfonate and p-toluenesulfonamide.

Marchese [34] has studied the effect of saccharine on the internal stress in nickel deposits obtained from a solution containing: $NiSO_4$, 33; $NiCl_2$, 45; and, H_3BO_3, 30 g/liter and has found that the sign of the stress is reversed. The author points out, however, that this effect is of brief duration.

Gorbunova and Popova [8] have studied the effect of surface-active substances in the electrodeposition of zinc and have shown that the internal stress in deposits obtained from a solution containing 450 g/liter $ZnSO_4 \cdot 7H_2O$ and 10 g/liter dextrine varies from 50 to 100 kg/cm², but does not diminish regularly with increasing deposit depth.

Much higher internal stresses are observed in zinc deposits obtained from a solution containing: $ZnSO_4 \cdot 7H_2O$, 215; potassium alum 30; and thiourea, 1.25 g/liter. Here, the internal stress rises sharply with increasing current density, reaches a maximum at 3.0 amp/dm², and then falls off as the current density is increased further.

It is interesting that the action of a surface-active substance on the internal stress is almost independent of the nature of the metal [31]. Thus, the internal stress in nickel is increased by the presence of 0.15 g/liter of thiourea, and reduced by the presence of 5 g/liter of naphthalenedisulfonic acid, comparison being made with the stress in deposits obtained from a pure electrolyte. Similar effects are observed in the deposition of copper in the presence

* The term "pitting" is used by electroplaters to designate marks left on the deposit by adhering hydrogen bubbles.

of these same additives (Fig. 65), addition of 0.005-0.025 g/liter thiourea increasing the internal stress in comparison with the stress in a deposit obtained from a pure electrolyte and the addition of 0.5 g/liter naphthalenedisulfonic acid reducing it. Thus, it is clear that an additive which will increase the internal stress in a nickel deposit will also increase the stress in a copper deposit.

Especial interest attaches to an understanding of the relation between the structure of the organic additive and its effect on the internal stress [31]. Figure 66 presents results on the effect of certain organic acids (valeric, propionic, and acetic) on the internal stress in copper deposits obtained from the standard sulfuric acid solution at a current density of 3.0 amp/dm². This figure makes it clear that the internal stress increases with the length of the hydrocarbon chain in the acid.

The combined action of two or more surface-active substances on the internal stress is a problem of great industrial importance. Figure 67 shows the combined effects of several pairs of additives in the deposition of copper. The additives employed here were: 0.005 g/liter thiourea + 0.5 g/liter naphthalenedisulfonic acid; 0.5 g/liter gelatine + 0.1 g/liter dextrine; 0.5 g/liter gelatine + 0.6 g/liter thiourea. The figure shows that the presence of two additives, one increasing and the other diminishing the internal stress, will give a stress which is lower than that arising from the action of the increasing additive alone. Used together, there is an increase in the effectiveness of two additives, each capable of giving rise to a tensile stress. The simultaneous presence of two additives, one giving rise to a tensile stress and the other a compressive stress, somewhat weakens the effectiveness of each.

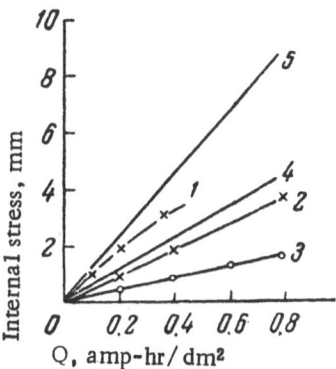

Fig. 64. The effect of current density on the internal stress in nickel deposits (Ise): 1) Electrolyte without additive, I_c = 3.0 amp/dm²; 2) electrolyte with added p-toluenesulfonamide (0.85 g/liter) and sodium benzylnaphthalenesulfonate (0.96 g/liter), I_c = 2.0 amp/dm²; 3) the same electrolyte with I_c = 1.0 amp/dm²; electrolyte with added sodium benzylnaphthalenesulfonate (0.96 g/liter), pH = 5.8, I_c = 1.0 amp/dm²; 5) the same electrolyte with I_c = 3.0 amp/dm².

Matulis and Valentelis [35] have studied the internal stress in copper deposits obtained in the presence of thiourea and Congo Red, or thiourea and benzylsulfoazonaphthylamine. The stress is high and the deposit quite brittle in the presence of thiourea alone, but the addition of Congo Red or benzylsulfoazonaphthylamine improves the quality markedly. Matulis and Valentelis explain these facts by pointing out that thiourea reacts with copper to form the complex $Cu(CSN_2H_4)_4^+$ which undergoes reduction on the cathode and thereby enhances the occlusion of thiourea in the deposit. The addition of Congo Red to the electrolyte leads to the formation of neutral molecules and thus reduces the likelihood that thiourea will enter the deposit.

A study of the effect of current density on the internal stress in deposits obtained in the presence of several additives has led to a curve which passes through a maximum [31] (Fig. 68). Compact uniform deposits are obtained in the neighborhood of this maximum point. A comparison of the curve obtained with thiourea alone and the curve obtained with thiourea and naphthalenedisulfonic acid together shows a change in both form and maximum stress, the second curve being flatter than the first. This is a very important fact since it makes possible a widening of the range of current densities for the preparation of high quality technical plates.

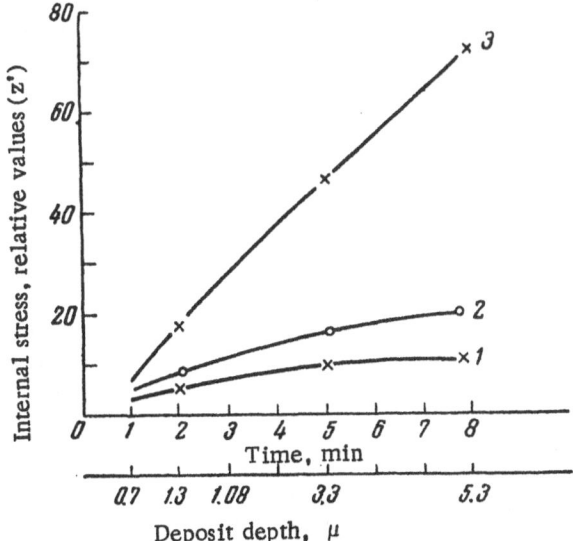

Fig. 65. The relation between depth and internal stress in copper deposits prepared in the presence of surface-active substances (Vagramyan, Petrova): 1) In the presence of 0.5 g/liter naphthalenedisulfonic acid; 2) electrolyte without additive; 3) in the presence of 0.015 g/liter thiourea. Deflection in millimeters equal to $z' \cdot 8.16 \cdot 10^{-2}$.

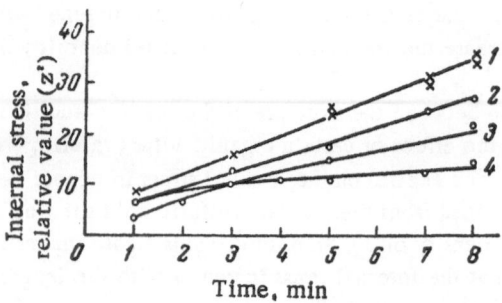

Fig. 66. The time variation of the internal stress in copper deposits and its relation to the structure of the organic additive (Tsareva, Solokhina, Kudryatsev, Vagramyan): 1) 0.03 M valeric acid; 2) 0.03 M propionic acid; 3) electrolyte without additive; 4) 0.03 M acetic acid. Deflection in millimeters equal to $z' \cdot 8.16 \cdot 10^{-2}$.

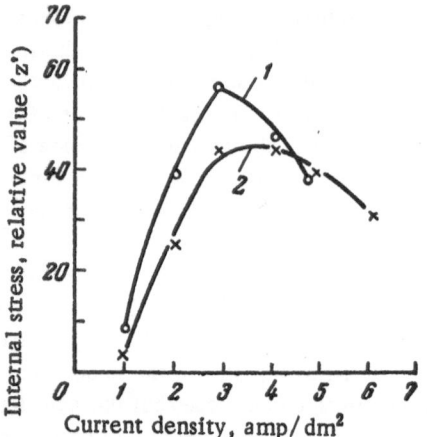

Fig. 68. Combined effect of several additives on the internal stress in deposits obtained at various current densities (Tsareva, Solokhina, Kudryatsev, Vagramyan): 1) With 0.005 g/liter added thiourea; 2) with 0.005 g/liter added thiourea + 0.5 g/liter naphthalenedisulfonic acid. Deflection in millimeters equal to $z' \cdot 8.16 \cdot 10^{-2}$.

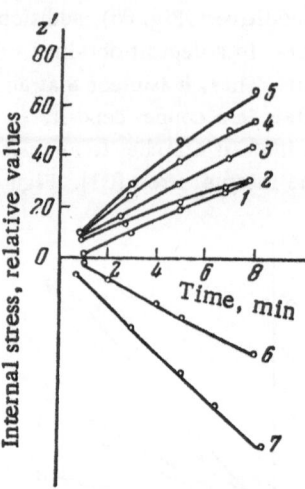

Fig. 67. Combined and separate effects of surface-active substances on the internal stress (Tsareva, Solokhina, Kudryavtsev, Vagramyan): 1) 0.005 g/liter thiourea + 5 g/liter freshly prepared naphthalenedisulfonic acid; 2) 0.2 g/liter dextrine; 3) 0.005 g/liter thiourea + 0.5 g/liter aged naphthelenedisulfonic acid; 4) 0.5 g/liter thiourea; 5) 0.5 g/liter gelatine + 0.1 g/liter dextrine; 6) 0.6 g/liter thiourea + 0.5 g/liter gelatine; 7) 0.6 g/liter thiourea. Deflection in millimeters equal to $z' \cdot 8.16 \cdot 10^{-2}$.

THE EFFECT OF FLUCTUATING CURRENT

There are many cases in which the internal stress can be reduced considerably by carrying out deposition with a fluctuating current. Two procedures are available for this purpose: 1) An alternating current can be superimposed on the direct current; 2) a reversing current can be employed.

Various closely related electrical circuits are used for carrying out deposition under superimposed direct and alternating currents.

Studies of deposits obtained by superimposing alternating currents of various frequencies on a direct current have shown the internal stress to be essentially constant above 500 cycles. Most work has therefore been carried out at standard frequencies, 50 cycles in the USSR, 60 cycles in the USA [34]. The internal stress is practically unaltered when the ratio of alternating to direct current (I_{alt}/I_{dir}) is less than unity. Table 20 gives certain data of Marchese on the internal stress in deposits obtained at various values of pH and various ratios of superimposed alternating and direct currents.

This table shows that the internal stress is increased by adding hydrogen peroxide to the electrolyte to eliminate pitting. For this reason, Marchese has suggested replacement of the peroxide by some other agent.

At pH 2.5, the greatest reduction in internal stress is observed with a 7 : 1 ratio of alternating to direct current (Fig. 69). The effect of superimposed alternating and direct currents has also been studied by Kohlschütter and Jakober [27], and Barklie and Davies [3].

TABLE 20. The Effect of Superimposed Alternating and Direct Currents on the Internal Stress in Nickel Deposits

Current density, amp/dm²	Deposit depth, μ	Current direct, amp	alternating, amp	Direct current potential, v	Alternating current potential, v	Potential ratio	Cathode displacement, relative values	Internal stress, kg/cm²	Notes
5.4	0.0013	1.0	0.0	1.6	0.0	0.0	0.0107	958	pH 2.0; 63°; 0.22 cm³/liter per day H_2O_2
5.4	0.0010	1.0	0.66	1.6	0.0	0.0	0.0100	854	pH 2.0
5.4	0.0009	1.0	1.08	1.55	0.50	0.3	0.0073	650	pH 2.0
5.4	0.0008	1.0	1.43	1.50	1.26	0.9	0.0070	589	pH 2.0
5.4	0.0010	1.0	2.43	1.50	2.60	1.7	0.0060	510	pH 2.0
5.4	0.0090	1.0	3.36	1.45	4.07	2.8	0.0038	321	pH 2.0
5.4	0.0090	1.0	3.36	1.45	4.07	2.8	0.0059	496	pH 2.0
2.7	0.0010	0.5	3.5	0.8	4.7	5.9	0.0058	496	pH 2.0
2.1	0.0010	0.42	1.07	0.8	1.2	1.5	0.0055	469	pH 2.0
4.3	0.0010	0.85	1.8	1.3	2.0	1.5	0.0058	496	pH 2.0
8.7	0.0010	1.65	3.0	2.1	3.3	1.6	0.0050	426	pH 2.0
5.4	0.0010	1.0	0.0	2.2	0.0	0.0	0.0223	1930	pH 5.0; 63°
5.4	0.0011	1.0	0.0	2.2	0.0	0.0	0.0196	1700	—
5.4	0.0010	1.0	0.0	2.2	0.0	0.0	0.0108	924	pH 4.5; 63°
5.4	0.0010	1.0	0.66	1.7	0.7	0.4	0.0123	1040	—
5.4	0.0010	1.0	1.1	1.8	0.3	0.17	0.0073	621	pH 4.5
5.4	0.0010	1.0	1.4	1.8	1.0	0.55	0.0051	433	
5.4	0.0010	1.0	2.4	1.6	2.5	1.6	0.0050	426	
5.4	0.0008	1.0	3.3	1.7	3.7	2.2	0.0059	490	
2.1	0.0010	0.4	1.0	0.8	1.0	1.3	0.0050	426	
4.3	0.0010	0.9	1.8	1.3	1.8	1.4	0.0040	341	
8.7	0.0010	1.7	3.4	2.2	3.3	1.5	0.0046	392	
5.4	0.0008	—	—	1.5	0.0	0.0	0.0088	727	
5.4	0.0010	—	—	1.5	1.0	0.7	0.0043	366	pH 2.5
5.4	0.0010	—	—	1.4	2.0	1.4	0.0054	462	
5.4	0.0010	—	—	1.3	3.0	2.3	0.0034	294	
5.4	0.0010	—	—	1.3	4.0	3.1	0.0029	245	
5.4	0.0010	—	—	1.3	6.0	4.6	0.0039	336	
5.4	0.0010	—	—	1.3	8.0	6.2	0.0033	280	
5.4	0.0010	—	—	1.2	10.0	8.3	0.0037	315	

The effect of a reversing current has been studied by Khonikevich, Fedot'ev, Rykova, and others.

Khonikevich and Fedot'ev have investigated the internal stress in copper deposits obtained with a 10 sec cathodic polarization and a 1 sec anodic polarization, i.e., a 10 to 1 ratio of cathodic to anodic polarization. They have shown that the passage from direct to reversing current will alter the stress considerably without changing the stress—

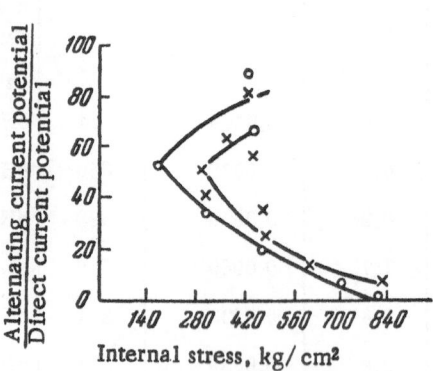

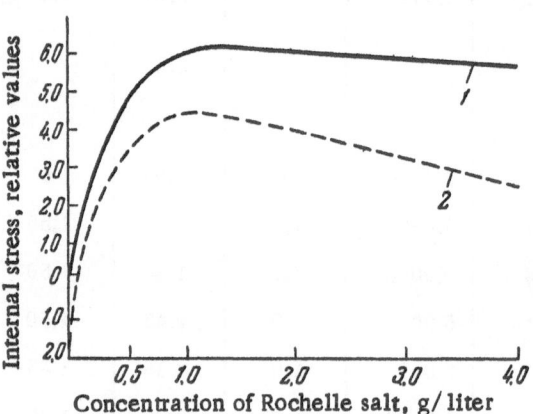

Fig. 69. The effect of superimposed alternating and direct currents on the internal stress (Marchese): $\bigcirc - 60$-cycle alternating current; $\times - 400$-cycle alternating current; $I_c = 5.4$ amp/dm^2, pH = 2.5; t = 63°.

Fig. 70. The relation between the concentration of Rochelle salt and the internal stress in copper deposits (Khonikevich, Fedot'ev): 1) Internal stress without superimposed alternating current; 2) internal stress with a 1 : 10 current reversal.

additive concentration relation. Deposits obtained under reversing currents show higher stress in the case of tension and lower stress in the case of compression, than do deposits obtained under direct current (Fig. 70).

Rykova has studied the effect of a reversing current on the internal stress in chromium deposits [1] obtained at various temperatures (50, 55, 65, and 70°) and various current densities (30, 50, and 100 amp/dm^2), the time between reversals ranging from 1 to 10 min. It was noted in this work that the stress was at a minimum in chromium deposits laid down in layers each less than 0.25 μ in depth. Increasing the depth of the individual layer to 5 μ led to a rise in the internal stress. There was no crack pattern with a reversal time of 3-5 min, but a clear-cut pattern developed when the layer depth was increased still further to 5-8 μ.

THE EFFECT OF AN INTERMITTENT CURRENT

Electrodeposition under an intermittent current involves passivation of the growing crystals and a marked alternation in the deposit structure. The internal stress in chromium deposits obtained with intermittent current has been studied by Rykova.

Electrolysis was continued for 1, 2, 3, 4, 5, or 10 min in these experiments, with an 1 min interruption period. The internal stress was reduced in each case, the diminution being largest for a 4-5 min electrolysis with a 1 min interruption period.

LITERATURE

1. Collection, Studies on the Corrosion of Metals under Stress [in Russian] (Mashgiz, Moscow, 1953).
2. M. L. Pertsovskii, Porous Chrome Plating [in Russian] (Mashgiz, Moscow-Sverdlovsk).
3. R. H. D. Barklie and J. J. Davies, Engineer 150, 670 (1930).
4. A. L. Rotinyan and E. S. Kozich, Zhur. Priklad. Khim. 31, 424 (1958).
5. N. P. Fedot'ev and Yu. M. Pozin, Zhur. Priklad. Khim. 31, 419 (1958).
6. V. S. Ioffe, Uspekhi Khim. 13, 50 (1944).
7. A. A. Khonikevich and N. P. Fedot'ev, Trudy Leningrad, Tekhnol. Inst. im. Lenosoveta, 40 (Goskhimizdat, Leningrad, 1957) p. 133.
8. K. M. Gorbunova and O. S. Popova, Zhur. Fiz. Khim. 30, 269 (1956).

9. V. V. Ostroumov, Zhur. Fiz. Khim. 31, 1812 (1957); Zhur. Priklad. Khim. 31, 402 (1958).

10. G. K. Potapov and A. T. Sanzharovskii, Zhur. Fiz. Khim. 32, 1416 (1958).

11. P. P. Stranskii and R. Kaishev, Uspekhi Fiz. Nauk 21, 408 (1939).

12. G. J. Finch, H. Wilman, and L. Yang, Disc. Faraday Soc. 144, 1125 (1947).

13. P. D. Dankov, Transactions, Second Conference on the Corrosion of Metals [in Russian] 2 (Acad. Sci. USSR Press, Moscow, 1943) p. 121.

14. G. Bliznakov, Ezhegod. Sofiisk. Un-ta., 2, Khim. ("Nauk i Iskusstvo" Press, Sofia, 1956) 49, p. 65.

15. C. Marie and Thon, J. Chem. Phys. 29, 11 (1932); Compt. rend. 193, 31 (1931).

16. D. T. Hurd, An Introduction to the Chemistry of the Hydrides (N. Y.-London, 1952).

17. A. Brenner and S. Senderoff, J. Res. Nat. Bur. Stand. 42, 89 (1949).

18. J. B. Kushner, Metal Finish. 56, 82 (1958).

19. A. T. Vagramyan and Yu. S. Tsareva-Petrova, Zhur. Fiz. Khim. 29, 185 (1955); A. T. Samartsev and Yu. V. Lyzlov, Zhur. Fiz. Khim. 29, 374 (1955); T. P. Hoar and D. Z. Arrowsmith, J. Electroplat. and Metal Finish. 10, 141 (1957).

20. J. B. Kushner, Metal Finish. 56, 52 (1958).

21. M. A. Shluger, The Theory and Practice of Electrolytic Chrome Plating [in Russian] (Acad. Sci. USSR Press, Moscow, 1957) p. 147.

22. L. A. Glikman, N. P. Fedot'ev, and A. P. Chernova, Zavodskaya Lab. 17, 1126 (1951).

23. A. T. Vagramyan and Z. A. Solov'eva, Methods of Investigating the Electrodeposition of Metals [in Russian] (Acad. Sci. USSR Press, Moscow, 1955).

24. N. P. Fedot'ev and E. G. Kruglova, Zhur. Priklad. Khim. 28, 273 (1955).

25. G. Stony, Proc. Roy. Soc. 82, 172 (1909).

26. H. Stager, Helv. chim. acta 3, 518, 614 (1920).

27. V. Kolschutter and F. Jakober, Z. Elektrochem. 33, 220 (1927).

28. E. A. Vuilleumier, Trans. Amer. Electrochem. Soc. 42, 99 (1922).

29. V. S. Ioffe, Dissertation, GIPKh.

30. G. S. Vozdvizhenskii, Zhur. Priklad. Khim. 20, 1171 (1947).

31. Yu. S. Tsareva, V. G. Solokhina, N. T. Kudryavtsev, and A. T. Vagramyan, Zhur. Fiz. Khim. 29, 166 (1955).

32. W. Rosenheim, Z. Metallkunde 22, 74 (1930).

33. N. Ise, J. Electrochem. Soc. Japan 26, 18 (1958).

34. V. J. Marchese, J. Electrochem. Soc. 99, 39 (1952).

35. Yu. Yu. Matulis and L. Yu. Valentelis, Trudy Akad. Nauk Lit. SSR, seriya B, 3, 17 (1957).

Chapter X

THE ORIGIN OF THE INTERNAL STRESS

There are numerous factors which can give rise to an internal stress, although it is usually true that just one of these is responsible for the stress in each particular case. These factors are [1]:

1. Alteration in the lattice parameters.
2. Alteration of the distance between the deposit crystals during deposition.
3. Enlargement of the deposit crystals by consolidation of smaller crystals.
4. Compound formation and alteration in volume as a result of reaction between the metal and foreign substances.

THE ALTERATION IN LATTICE PARAMETERS

Electrodeposition produces an unstable crystal lattice of metal atoms with distorted lattice parameters which tends to pass over into a stable state, thereby giving rise to an internal stress. The following factors can lead to a distortion of the lattice parameters.

a) The high potential drop in the electric double layer increases the mean thermal energy of the discharging particle as it enters the crystal lattice and thereby increases the lattice parameters. The Lentz-Joule thermal effect which is observed at the electrode—electrolyte inferface during passage of current through the cell is considerably larger. With ordinary current densities, the potential drop is of the order of 0.01-0.1 v/cm in the electrolyte, and 10^7-10^8 v/cm in the double layer, the resistance at the electrode—solution interface being 8-9 orders higher than the resistance of the electrolyte. Since the bath, itself, is heated by several degrees during electrolysis, there must be a colossal temperature rise in the double layer where the resistance is high and the dissipation of thermal energy is poor. No methods exist at the present time for measuring thermal effects in the electric double layer, but certain data on the electrodeposition of powders indicate that the temperatures in question are, indeed, rather high [2]. The dissipation of heat from the double layer during formation of electrolytic powders must be poor because the particles make imperfect contact with one another and the convection currents are weak. Here, an ordinary thermometer will disclose an elevation of temperature in the thicker electrolyte layers surrounding the electrode. This, it can be concluded that there is a profound alteration in the ion temperature at the time of reduction which must certainly affect the state of the crystal lattice.

It also follows that high internal stress should be observed in deposits formed at high overvoltage. The data of Table 16 confirm this conclusion.

b) Distortion of the crystal lattice can also arise from atomic occlusion of hydrogen or other foreign substances. The distortion observed in such cases is due to the fact that the force field of the occluded atom is not the same as the force field of the metal atom and a tensile or compressive stress is set up. The work of Rosenheim [3] has proven that the lattice parameters of copper are increased by occlusion of aluminum or zinc and reduced by occlusion of nickel, while the work of Vagramyan and Tsareva [1] has shown that the internal stress in copper is increased by deposition in the presence of zinc or aluminum salts and diminished by deposition in the presence of nickel salts. These facts confirm the previously assumed existence of a relation between the character of the internal stress and the alteration in crystal lattice parameters.

Ioffe [4] considers that the lattice parameters alter as crystal growth proceeds in the electrodeposit. Here, the assumption is made that the lattice parameters of a surface layer are not the same as those of an internal layer of the crystal. The lattice parameters must, of course, depend on the interaction between neighboring ions, or atoms, in the lattice. For this reason, the parameters of a surface layer of a homopolar crystal must be smaller than the parameters of an internal layer, the inner atoms being attracted by more atoms than the outer. The situation is just the opposite in ionic crystals where the inner ions are repelled by a greater number of neighbors than the outer and the higher parameters are found in the surface layer.

Progressive deposition converts the surface layers of the lattice into internal layers and tends, according to Ioffe, to alter the lattice parameters, thus giving rise to a tensile or compressive stress. It is Ioffe's opinion that this gives an explanation of the fact that the stress is higher in fine-grained crystalline deposits than in coarse, since the finer the crystals, the larger the fraction of volume represented by external layers, and the greater the tendency of the lattice parameters to alter as deposition proceeds.

THE ALTERATION IN THE DISTANCE BETWEEN CRYSTALS

It is well known that the electrolytic deposit occludes a considerable proportion of the foreign particles which are present in the electrolyte, either through adsorption (of surface-active substances), migration under the action of the electric field, chemical union with the metal, mechanical processes, or still other means. Some of these oc-cluded particles are taken into the crystal lattice, while others are distributed along the grain boundaries. Various factors can cause an alteration in the distance between crystals. First, there is the fact that the molecules of surface-active materials are situated in the electric double layer and are deformed by the action of this field, being stretched in one direction or another. Such deformation is especially marked in organic compounds. Crystal growth and the accompanying displacement of the double layer into the electrolyte eliminate the action of this field on the occlud-ed molecules and the latter push the crystals apart in striving to acquire normal form.

A second factor tending to alter the crystal separation is the irregular distribution of occluded foreign parti-cles between crystals. Diffusion leads to a redistribution of these particles in the deposit and gives rise to an in-ternal stress.

Foerster [5] has given an analogous explanation of the effect of occluded hydrogen on the internal stress, point-ing out that the stress arises from the irregular distribution of this gas throughout the deposit (the first layers of a nickel deposit are richer in hydrogen than are subsequent layers). He has not, however, advanced a detailed mechan-ism for the effect of this irregular hydrogen distribution on the internal stress.

An irregular distribution of foreign particles throughout the deposit is also met in the case of a surface-active substance whose concentration diminishes steadily as electrolysis proceeds.

THE ALTERATION IN CRYSTAL DIMENSIONS

a) The irreversible deposition of a metal can lead to the appearance of a finely crystalline mass which will undergo subsequent aggregation in an attempt to reduce the surface energy and, thereby, give rise to an internal stress. An increase in temperature accelerates this process by facilitating the interchange of atoms, while the ad-sorption of contaminants on the grain boundaries retards it.

Kohlschütter [6] considers that the first metal to be brought down by electrolysis is in a highly dispersed form of high, free energy. This metal is metastable and passes into a more stable state by crystal aggregation, the ac-companying reduction in volume giving rise to a compressive stress. A somewhat different point of view has been advanced by Nemonov [7] who considers the stability of any one lattice type to be fixed by the temperature and crystal dimensions. He claims, for example, that while energy considerations indicate the hexagonal form to be the most suitable for those chromium crystals of small diameter which are first laid down during deposition, the body-centered cubic lattice with lower specific volume becomes the most suitable form as growth proceeds. The hexa-gonal lattice has lower free energy when the crystal radius is small and is, therefore, more stable than the cubic, but as the crystal radius increases this situation changes and the cubic lattice becomes the more stable. Thus, crys-tal growth is accompanied by a passage from hexagonal to cubic lattice types which gives rise to an alteration in the volume of the deposit and an internal stress.

b) Wyllie [8] considers that hydrogen separates out simultaneously with the metal and is largely occluded in the deposit, giving rise to a pressure which can, on occasion, be so high to lead to deformation and crystal disinte-gration. The effect, here, is similar to that which is observed in the drawing and rolling of metals and results in an increase in the number of orientated crystals. It is for this reason that x-ray studies show deposit orientation, espe-cially in those metals whose deposition is accompanied by an evolution of hydrogen. Thus, Wyllie believes the in-ternal stress to be related to the occlusion of hydrogen during deposition, a point which he supports through data in-dicating the existence of high compressive stress in metals whose deposition is accompanied by hydrogen evolution. The occlusion of hydrogen and the internal stress can be reduced by elevating the temperature or by superposing an alternating current.

The interaction between metal and occlusions can lead to the formation of a chemical compound whose volume is either greater or less than that of the components; an internal stress can arise in the deposit as a result.

Either one or several of these factors can predominate under the conditions actually prevailing during deposition.

Thus, an understanding of the origin of the internal stress can facilitate a selection of the conditions required for obtaining metal deposits with predetermined internal stress.

LITERATURE

1. A. T. Vagramyan and Yu. S. Tsareva, Zhur. Fiz. Khim. 29, 185 (1955).
2. D. N. Gritsan and A. M. Bulgakova, Zhur. Fiz. Khim. 28, 258, 337 (1954); 29, 448, 649 (1955); 31, 1943 (1957); Doklady Akad. Nauk SSSR, 100, 1111 (1955).
3. W. Rosenheim, Z. Metallkunde 22, 74 (1930).
4. V. S. Ioffe, Uspekhi Khim. 13, 144 (1944).
5. F. Foerster, Elektrochemie wässeriger Lösungen (Leipzig, 1922).
6. V. Kolschütter and F. Jakober, Z. Elektrochem. 33, 220 (1927).
7. A. S. Nemnonov, Zhur. Tekh. Fiz. 18, 238 (1948).
8. M. R. Wyllie, J. Chem. Phys. 16, 52 (1948); W. Hume-Rothery and M. R. Wyllie, Proc. Roy. Soc. A. 182, 131 (1943).

Chapter XI

THE INTERNAL STRESS AND POROSITY OF THE ELECTROLYTIC DEPOSIT

The porosity is one of the principal, determining factors in fixing the quality and corrosional stability of the electrolytic deposit. Study has shown that the porosity is closely related to the internal stress in the deposit.

One of the most important problems in the study of electrodeposition is that of accounting for the formation of pores in the metal deposit. The term "pore" is usually used to designate those micro- and macrocanals which lead from the surface of the deposit to the cathodic base metal. The determination of the number and dimensions of the pores is a very complex problem, since there is usually a wide distribution of pore diameters in the deposit. The presence or absence of pores in a given type of deposit will depend on the experimental conditions and the duration of the experiment, so that any evaluation of porosity can have only relative significance.

Pores can be separated arbitrarily into three groups on the basis of their diameters.

1. Macropores. The principal factor responsible for the formation of these pores is the irregularity in the electrode surface resulting from occlusion of foreign metallic and nonmetallic substances, oxides, surface-active materials, etc.

2. Micropores. The formation of such pores is controlled by the deposit structure which depends, in turn, on the conditions of electrolysis.

3. Channel pores or crack nets. These pores result from the high internal stress in the deposit.

The variability in pore forms and diameters and the lack of exact methods for exposing the pores are the principal sources of difficulty in any study of the porosity of the electrolytic deposit.

Relatively little work has been done on porosity and much of this has been incidental to a study of other problems in the electrodeposition of metals.

Most studies of porosity have been carried out on technically important electrolytic deposits of nickel and chromium on various types of steel.

The pores in these deposits can be exposed by the use of standard ferrocyanide reagent containing 10 g/ liter $K_3Fe(CN)_6$ and 20 g/ liter NaCl.

The essential feature of this procedure is that the reagent penetrates the pores to the base metal, reacts with this base, and the reaction products then pass back through the pores to form blue spots on the surface. The number of pores is assumed to be equal to the number of spots.

The accuracy of determination of the porosity is generally improved by carrying out this count under the microscope.

THE EFFECT OF THE BASE AND THE PRELIMINARY TREATMENT

It has been shown above that the base affects both the structure and internal stress of the deposit, and it follows that it will also have a marked influence on the porosity [1-3]. Studies on nickel cathodes have indicated that the number of pores in the deposit will increase with the roughness of the electrode surface and diminish as this surface is smoothed out [4].

The best results are obtained if the surface is first subjected to mechanical treatment, then carefully washed, and finally degreased by being rubbed with soda until it is entirely wet by water.

The pore number is quite high in metal deposits which have nonuniform structure because of the presence of various occlusions. Thus porous deposits can result from the presence of nonmetallic occlusions (which enter the cathode surface during mechanical treatment), foreign metals and alloys, fats and surface-active substances (which remain after inadequate degreasing), and slags.

The occlusion of metals with low hydrogen overvoltage affects the porosity quite markedly. The sectors of the cathode having low hydrogen overvoltage are sectors of high metal overvoltage [5] and on them, there is an intensive evolution of hydrogen and a retarded deposition of metal. The intensive evolution of hydrogen on these sectors leads to the formation of craterlike depressions, each with a canal leading to the base metal at its center. Continuous evolution of hydrogen bubbles gives rise to an appreciable furrowing of the cathode surface.

A microsection prepared by cutting the specimen through one of these craterlike depressions shows the large pore to have been formed at an occlusion.

It should be noted that pores can form, not only on sectors with low hydrogen overvoltage, but also on sectors which are poorly wet by the electrolyte because of inadequate degreasing.

It has been shown in a number of papers (see especially, [6]) that the strength of hydrogen bubble adhesion and the bubble diameter at the instant of break-off depend on the so-called contact angle θ which is defined in terms of the surface tensions at the solution–gas (σ_{23}), solution–metal (σ_{21}), and metal–gas (σ_{13}) interfaces through $\theta = \dfrac{\sigma_{13} - \sigma_{12}}{\sigma_{23}}$. The rate of bubble break-off increases as this angle diminishes. Thus Kabanov and Fainglauz [7] have shown that bubbles for which θ = 15° break-off 50 times faster than bubbles for which θ = 50°. The Neumann Rule predicts that the contact angle will be closely dependent on the presence of surface-active substances in the electrolyte.

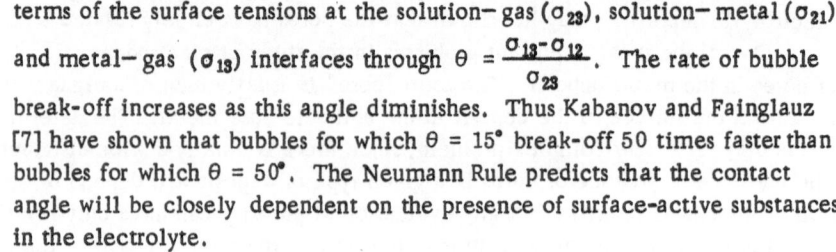

Fig. 71. The relation between electrode potential and contact angle. (Gorodetskaya, Kabanov).

The contact angle varies with the electrode potential and the curve covering this relation passes through a maximum [8] (Fig. 71).

Thus the adhesion time and the bubble diameter depend on both the concentration of surface-active substances and the cathodic current density. The separation of metal is retarded and pores are formed on those sectors of the cathode to which hydrogen bubbles adhere for an extended period [9].

THE EFFECT OF THE STRUCTURE AND DEPTH OF THE DEPOSIT

The dimensions of the crystals clearly determine the degree of coverage of the electrode surface by a fixed amount of deposited metal. A coarse crystalline deposit will cover only a small fraction of the electrode surface, whereas the same amount of metal in finely crystalline form will deposit uniformly over the entire area.

The amount of metal required for so covering the entire surface of the cathode as to obtain a nonporous deposit will be much greater for large crystals than for small.

Thus, to each crystal diameter there will correspond a certain minimum amount of metal which will just cover the entire electrode surface. This can be designated provisionally as the "critical" amount. The "critical" depth for a nonporous deposit will vary with the crystal diameter, since the mean deposit depth is proportional to the amount of metal laid down on the cathode. Thus the minimum depth required for a nonporous deposit must increase with the crystal diameter.

Curves covering the relation between pore number and deposit depth for orientated, densely packed crystals of various dimensions will show sharp bends corresponding to complete coverage of the surface with a nonporous plate.

The preparation of a thin, nonporous plate necessitates selection of those conditions of electrolysis which will give a uniform, finely crystalline deposit.

The crystals obtained in actual electrolysis are unoriented and haphazardly distributed for the most part; the relation between crystal diameter and minimum depth is more complex, and the break on the pore number, deposit depth curve corresponding to complete coverage is marked less clearly. The minimum depth required for forming a nonporous deposit with disordered distribution will be much greater than that calculated by assuming dense crystal packing.

Figure 72 [4] presents the results obtained in a study of the relation between porosity and depth in nickel deposits. Here, nickel was deposited onto iron from a solution containing: $NiSO_4 \cdot 7H_2O$, 166; H_3BO_3, 30; $Na_2SO_4 \cdot 10H_2O$, 20; $MgSO_4 \cdot 7H_2O$, 20; and, KCl, 2 g/liter working at a current density of 2 amp/dm^2 and a temperature of 17-20°.

These porosity curves have been constructed around mean depth values obtained by dividing the increase in mass by the total surface area of the electrode. Curve 1 shows that the pore number first falls rapidly as the depth increases, levels off at depths of the order of 15-17 μ, and then remains essentially constant as the depth increases further. It is impossible to obtain completely nonporous deposits, even by going to 355 μ.

The existence of a great number of pores on the curve segment corresponding to deposit depths ranging up to 17 μ is due to the fact that the depth has not yet reached the "critical" value, while certain other factors account for the existence of pores on the segment corresponding to depths in excess of 17 μ. These figures are in rather good agreement with the data of other authors [10]. The relation between pore number and deposit depth is the same for copper and iron bases (Curves 2 and 3).

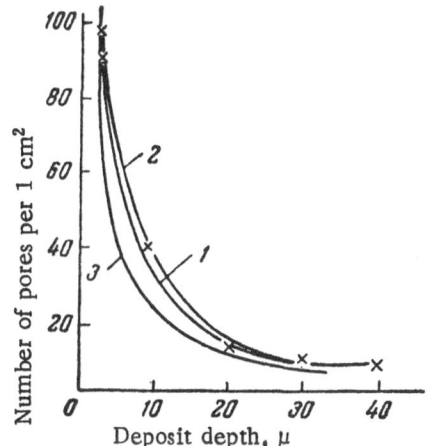

Fig. 72. The relation between depth and pore number in nickel deposits: 1) Iron base (Vagramyan, Sutagina); 2) copper base; 3) iron base (Rotinyan, Fedot'ev, Mishchenkova, and Te Tu Huo).

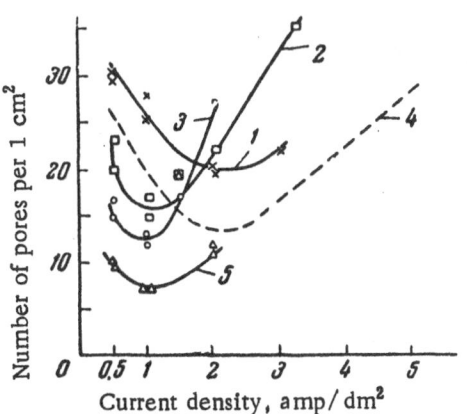

Fig. 73. The relation between current density and pore number in nickel deposits obtained with various added surface-active substances (Vagramyan, Sutyagina): 1) 0.15 g/liter p-cresol; 2) 8 cm³/liter formaldehyde; 3) 8 cm³/liter acetone; 4) without addition; 5) 3 g/liter 2.6-2.7 naphthalenedisulfonic acid. Composition of electrolyte: $NiSO_4 \cdot 7H_2O$, 210 g/liter; H_3BO_3, 30 g/liter; NaCl, 3 g/liter; and, NaF, 6 g/liter.

THE EFFECT OF CURRENT DENSITY

The curves of Fig. 73 cover the relation between the current density and the pore number in nickel deposits [9] laid on an iron base in 17-20 μ plates.

Curve 4 of this figure shows that the current density, pore number curve passes through a minimum which corresponds to 2 amp/dm² at the temperature and electrolyte composition in question.

The appearance of this minimum accounts for the fact that dense, uniform, metallic deposits are obtained in a definite interval of current densities. The increase in porosity at higher and lower current densities reflects the deterioration in quality of the deposit.

The pore number rises in nickel deposits with increasing current density because of a deterioration of quality resulting from intensive hydrogen evolution [8] and alkalization of the layer around the electrode. Internal stress studies have shown that the curve covering the stress 1 current density relation also passes through a maximum the deposit obtained at this point having highest uniformity and density of packing [11]. The fall-off of internal stress at lower and higher current densities is due to a reduction in the quality of the deposit. Thus, comparison of internal stress, porosity and internal stress, current density curves should be made for deposits of maximum density and highest quality, minimum porosity on the one curve corresponding the maximum stress on the other.

Rotinyan, Fedot'ev, Mishchenkova, and Te Tu Huo have studied the porosity of nickel deposits prepared from several different electrolytes [10] and have concluded that minimum porosity is obtained at density of 1.5 amp/dm².

THE EFFECT OF SURFACE-ACTIVE SUBSTANCES

Vagramyan and Sutuagina [4] have studied the effect of various surface substances (acetone, formaldehyde, p-cresol, etc.) on the porosity of nickel.

The results on the relation between pore number and current density in the presence of various additives are shown in Fig. 73.

This figure makes it clear that the pore number, current density curve passes through a minimum even in the presence of surface-active additives. The minimum in the case of acetone, formaldehyde, or 2,6-2,7 naphthalenedisulfonic acid is displaced toward lower current densities in comparison with the minimum on the curve for the pure electrolyte. The deposit with minimum number of pores is that obtained by deposition from a solution containing naphthalenedisulfonic acid. The displacement of the minimum results from the fact that a dense uniform deposit is obtained at lower current densities after introduction of the additive into the electrolyte.

Loshkarev, Esin, and Sotnikova [12] have also shown that the current density required for obtaining a dense, uniform deposit is lowered by the introduction of a surface-active substance.

Thus, the displacement of the minimum on the pore number-current density curve toward lower current densities results from the fact that only there can a dense, uniform deposit be obtained.

The increase in pore number with current density in the presence of such additives as formaldehyde and acetone is due to the intensive reduction of these substances on certain sectors of the cathode, which has the same effect as an evolution of hydrogen. Studies made under similar conditions have shown that [11] the addition of naphthalenedisulfonic acid to the electrolyte will reduce the stress, while the addition of p-cresol will increase it (Fig. 74). The somewhat different curve forms of this figure probably arise from the fact that the porosity of the nickel was determined on steel and the internal stress measured on brass.

Figure 74 shows that those surface active substances which increase the porosity will also increase the internal stress, and vice versa.

THE EFFECT OF AN ALTERNATING OR REVERSING CURRENT ON THE INTERNAL STRESS AND POROSITY

Studies on the effect of an alternating current [1] have confirmed the existence of a certain parallelism between internal stress and porosity. Figure 75 presents data on the effect of an alternating current on the internal stress in deposits obtained from an electrolyte containing: $NiSO_4$, 250; $NaCl$, 5; and, H_3BO_3, 30 g/liter at pH 5.5, 20°, and $I_c = 2$ amp/dm^2. The alternating current density was varied from 1 to 2 amp/dm^2. This figure shows that

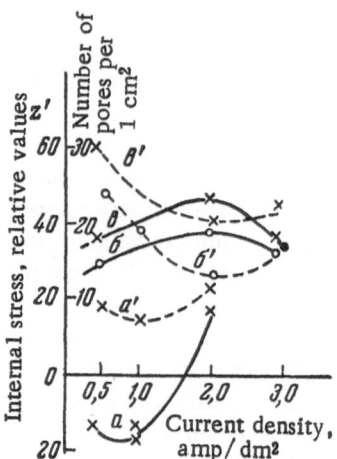

Fig. 74. The relation between porosity and internal stress in nickel deposits (Vagramyan, Petrova): a) Internal stress in deposits obtained in the presence of 3 g/liter 2,6-2,7 naphthalenedisulfonic acid; b) internal stress in deposits obtained without additive; c) internal stress in deposits obtained from an electrolyte containing 0.15 g/liter added p-cresol; a', b', c') corresponding porosity curves. Deflection in millimeters equal to z' · 8.16 · 10^{-2}.

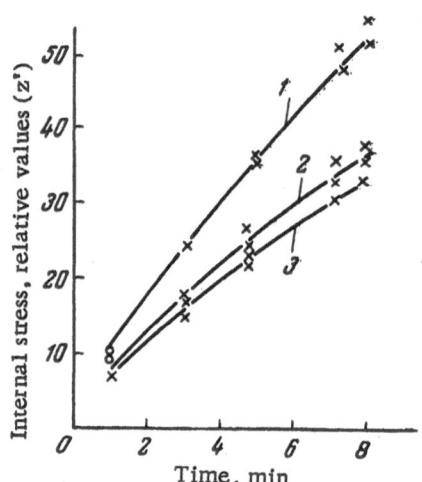

Fig. 75. The time variation of the internal stress in nickel deposits with superposed alternating current (Vagramyan, Petrova): 1) With superposed alternating current; 2) $I_{c(direct)} = 2.0$ amp/dm^2, $I_{c(alt)} = 1.0$ amp/dm^2; 3) $I_{c(direct)} = 2.0$ amp/dm^2, $I_{c(alt)} = 2.0$ amp/dm^2. Deflection in millimeters equal to z' · 8.16 · 10^{-2}.

the internal stress in the deposit is reduced by superposition of an alternating current. An investigation of the porosity carried out under analogous conditions [3] showed that the superposition of an alternating current of 1 amp/dm^2 on a direct current of 2 amp/dm^2 would lower the porosity from 14 pores/cm^2 to 8 pores/cm^2.

Bakhvalov and Rumyantsev [13] have studied the internal stress and porosity in electrolytic deposits obtained with a reversing current and have found that the deposits with lowest internal stress are also those of lowest porosity.

It follows from what has been said that the internal stress and porosity of dense deposits are affected similarly by surface-active substances and by alternating currents. This fact points to a common origin for the microporosity and internal stress of the deposit.

THE EFFECT OF TEMPERATURE AND ELECTROLYTE COMPOSITION

Rotinyan, Fedot'ev, Mishchenkova, and Te Tu Huo [10] have studied the effect of temperature and electrolyte composition on the porosity of nickel deposits. These studies have shown that the porosity falls off sharply up to 50° and then remains essentially constant with further elevation of temperature (Fig. 76). This type of relation is consistent with the data on the temperature variation of the internal stress.

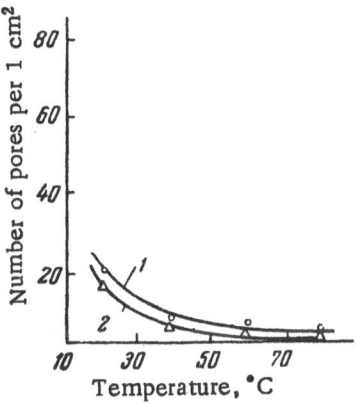

Fig. 76. The effect of temperature on the porosity of nickel deposits (Rotinyan, Fedot'ev, Mishchenkova, Te Tu Huo): 1) Without agitation; 2) with agitation.

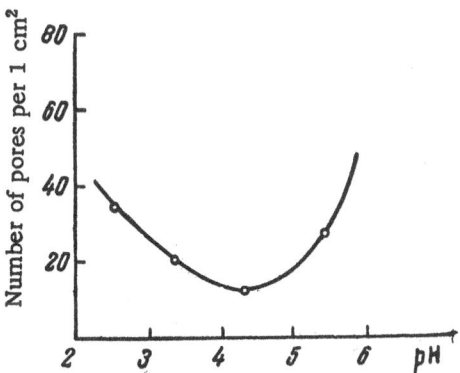

Fig. 77. The effect of the pH of the electrolyte on the porosity of nickel deposits. (Rotinyan, Fedot'ev, Mishchenkova, Te Tu Huo).

These same workers have studied the effect of concentration (nickel sulfate, sodium chloride, boric acid, sodium sulfate, and magnesium sulfate), pH of the electrolyte, and other factors, on the porosity.

This work has shown that the curve covering the relation between porosity and solution pH passes through a minimum in the area of pH 4-5 (Fig. 77). Thus minimum porosity coincides with minimum internal stress (Fig. 53).

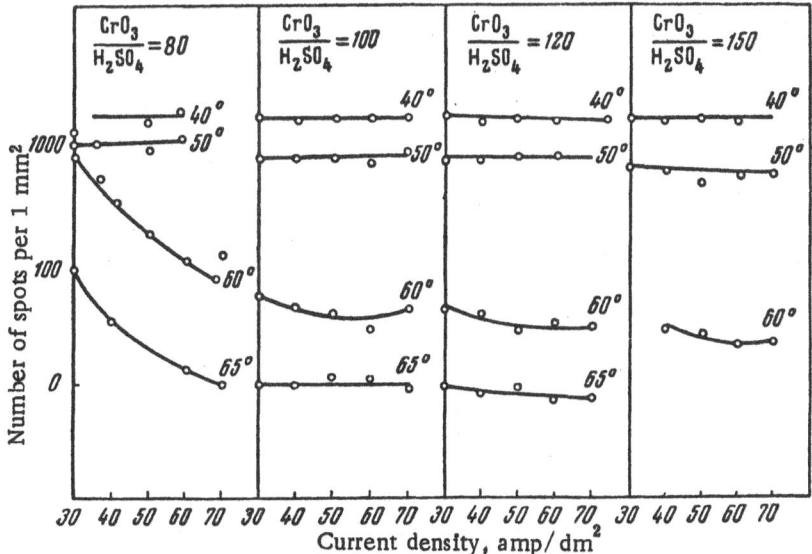

Fig. 78. The relation between the number of canal pores and the current density at various temperatures (Shluger).

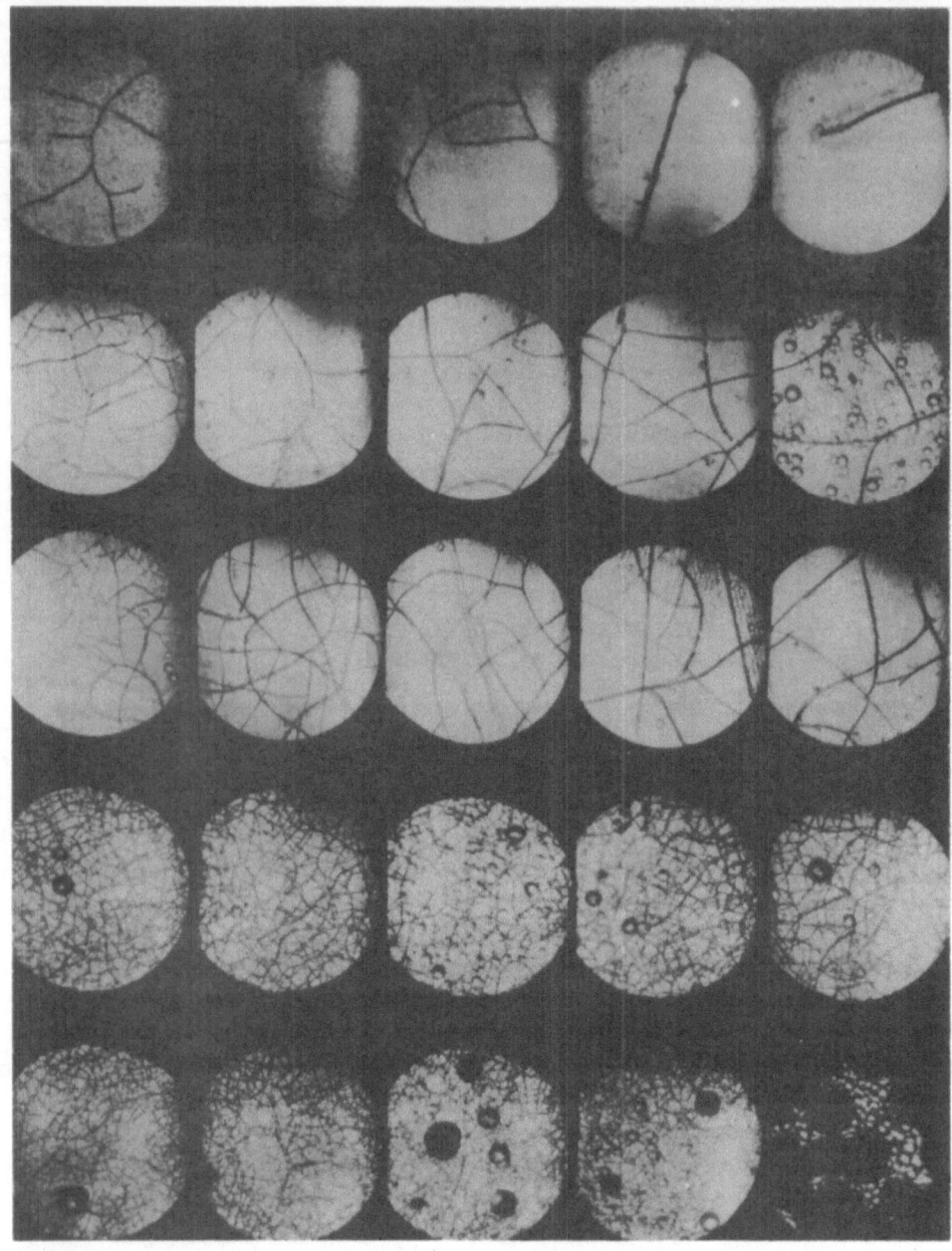

Fig. 79. Microphotographs of electrolytic chromium deposits obtained under various conditions (Shluger).

This thorough study of Rotinyan, Fedot'ev, Mishchenkova, and Te Tu Huo has proven that electrolytic nickel deposits of minimum porosity and maximum corrosional resistance are obtained from an electrolyte containing: $NiSO_4 \cdot 7H_2O$, 220; NaCl, 10; and, H_3BO_3, 30 g/liter working at pH 4.2-4.6, I_C = 1.5 amp/dm^2, 50°, and a depth of 20 μ.

CANAL PORES

It has been brought out above that the existence of high internal stress in the deposit will lead to rupture and the formation of crack systems, thus increasing the number of canal pores markedly. This kind of porosity is most frequently observed in chromium, palladium, and manganese. Chromium porosity is of especial interest, since this metal is widely used for wear resistant platings on the working parts of machinery. Pores are not always objectionable, since they can, in certain cases, hold a lubricant and spread it continuously and uniformly over the entire working surface. Porous chromium has many practical applications and a considerable amount of work has, therefore, been devoted to its study. Methods have been developed for bringing out the pore system of the deposit and the effect of the conditions of electrolysis on the number of such pores has been investigated. Anodic etching of the deposit is generally resorted to in order to bring out the system of canal pores. The canal type of porosity is usually estimated by microscopic count of the number of pores per unit surface area.

The graphs of Fig. 78 show the relation between current density and number of canal pores in porous chromium deposits obtained at various temperatures and various values of the ratio, chromic anhydride to sulfuric acid [14], the initial amount of chromic anhydride being 250 g/liter in each experiment.

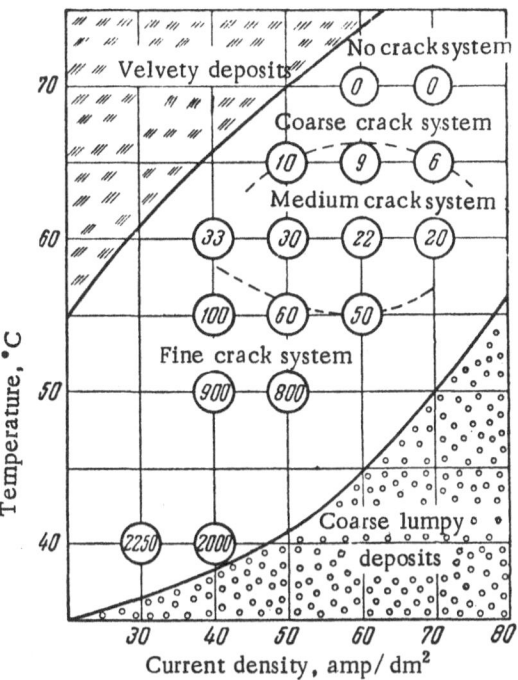

Fig. 80. Diagram showing the conditions required for preparing chromium deposits of various structures (Shluger). The figures in the circles characterize the porosity of the deposit and represent the number of spots formed on 1 mm² of surface by the crack system.

It follows from this figure that the porosity is strongly dependent on the temperature and decreases sharply as the temperature rises, the fall-off being especially marked in the 50-60° interval. The number of pores also diminishes with increasing current density, the fall-off being the sharper the higher the temperature. The porosity of the deposit cannot be fully characterized by the number of canal pores since these pores can vary in both depth and diameter.

Figure 79 shows microphotographs of porous chromium deposits obtained under various conditions. The pore number diminishes with increasing temperature and current density, but there is, at the same time, a change in the character (depth and diameter) of the porosity.

A diagram marking out the conditions required for producing porous chromium deposits is of much interest in preparing platings of this type [14] (Fig. 80). Such diagram contains three principal regions in which the conditions are favorable for the preparation of porous chromium. A deposit obtained at low temperature and high current density has a rough surface and is, from what has been said above, of no practical value. A deposit obtained at high temperature and low current density has a grayish blue sheen, especially after anodic etching, and is sometimes designated as "velvety." The porous deposits of greatest practical interest are obtained under the conditions marked out in the central part of the diagram.

These experimental results show that there is a relation between porosity and internal stress, porosity always increasing with stress regardless of the pore type.

LITERATURE

1. N. T. Kudryavtsev and G. Meshcherina, The Electrolytic Plating of Iron with Nonporous Nickel Films [in Russian] (Tsentr. In-t. Tekn.-Ekonom. Informatsii Soveta, NKTP, 1934).
2. A. W. Hothersall and R. A. Hammond, Trans. Amer. Electrochem. Soc. 73, 449 (1938).
3. M. I. Zil'berfarb, Zavodskaya Lab. 11, 1119 (1945).
4. A. A. Sutyagina and A. T. Vagramyan, Zhur. Priklad. Khim. 24, 945 (1951).
5. K. Frolich and F. Clark, Z. Elektrochem. 31, 649 (1925).
6. B. N. Kabanov and A. N. Frumkin, Zhur. Fiz. Khim. 4, 539 (1933).
7. B. N. Kabanov and E. Faingluz, Zhur. Fiz. Khim. 8, 795 (1936).
8. A. Gorodetskaya and B. N. Kabanov, Zhur. Fiz. Khim. 4, 529 (1933).
9. A. T. Vagramyan, Zhur. Fiz. Khim. 19, 305 (1945).
10. A. L. Rotinyan, N. P. Fedot'ev, E. E. Mishchenkova, and Te Tu Huo, Inform.-Tekhnich. Listok No. 74, Leningrad. Doma N.-T. Prop. (1956).
11. A. T. Vagramyan and Yu. S. Tsareva, Doklady Akad. Nauk SSSR 98, 807 (1954).
12. M. L. Loshkarev, O. Esin, and V. Sotnikova, Zhur. Obshchei Khim. 9, 1412 (1939).
13. G. T. Bakhvalov and N. V. Rumyantsev, Electrolytic Metal Plating with a Reversing Current [in Russian] (Moskovsk. Dom. N.-T. Prop. im. F. E. Dzerzhinskogo, 1957).
14. M. A. Shluger, Collection, Papers on the Theory and Practice of Electrolytic Chrome Plating [in Russian] (Acad. Sci. USSR Press, Moscow, 1957), p. 147.

Chapter XII

ADHESION AND THE INTERNAL STRESS
IN THE ELECROLYTIC DEPOSIT

The adhesion of the deposit to the base depends, in many cases, on the character and magnitude of the internal stress. The internal stress, adhesion, and quality of the deposit, all depend on the conditions of electrolysis. It is known that an increase in the current density can lead to the formation of a rough deposit and this, naturally, does not adhere as well as a compact deposit would. It frequently happens that internal stresses of considerable magnitude are set up in preparing dense electrolytic deposits and the adhesion with the base is, thereby, weakened. This can lead to a spontaneous peeling of the deposit from the base. Jacquet [1] has studied the effect of internal stress on adhesion and has shown that the adhesion of copper to nickel is reduced markedly when deposition is carried out in the presence of organic substances which give rise to an internal stress.

Table 21 gives certain data of Jacquet on the adhesion and internal stress in copper deposits laid down on a crudely cleaned nickel surface and the dependence of these factors on the previous history of the bath.

These figures show that the adhesion of deposit to base improves with an increase in the time of working the electrolyte. This same author has also shown that the internal stress in the deposit diminishes as the time of working the electrolyte increases.

It is quite likely that these results are due to a partial elimination of surface-active substances from the electrolyte by decomposition at the electrode during the working of the bath.

TABLE 21. The Relation between Internal Stress, Adhesion, and Time of Preliminary Working of the Electrolyte

Composition of electrolyte, g/ liter		Time of preliminary working of the bath, hours	Adhesion, g/ unit of surface	Internal stress, kg/ cm^2
CuSO$_4$ · 5H$_2$O	·H$_2$SO$_4$			
250	96	140	10,200	130
250	96	14	5700	250
250	48	14	550	500

A comparison of these figures shows that the adhesion of the deposit to the base becomes weaker as the internal stress increases. Excessive hydrogen occlusion frequently weakens the adhesion and sometimes leads to a swelling of the deposit.

Fedot'ev and Kruglova [2] have studied the effect of internal stress on adhesion in the case of the protection of silvered mirrors by copper films. It is known that a silver mirror is often protected from mechanical damage and interaction with chemical reagents by being covered with a compact, homogeneous copper film. The internal stress which arises during deposition of this film causes the silver to peel away. Fedot'ev and Kruglova have attempted to eliminate this defect by the use of such additives as sodium naphthalenedisulfonate, phenol, foramine, dextrine, and Rochelle salt and have shown that only the last of these substances can prevent peeling. The addition of 0.2 g/ liter of Rochelle salt will eliminate peeling completely, and at the same time, improve the quality of the deposit. The best results are obtained with an electrolyte containing $CuSO_4 \cdot 5H_2O$, 200; H_2SO_4, 2-5; and Rochelle salt, 1 g/liter working at I_C = 1-2 amp/dm².

Fedot'ev and Kruglova have also used the cathode bending method to study the relation between internal stress and concentration of Rochelle salt. The results of this study are presented in Fig. 81 where it is seen that the sing of the internal stress is altered by adding this salt to the acidic copper electrolyte. A stress which is tensile in the absence of the Rochelle salt becomes compressive on addition of small quantities of this compound.

It was pointed out above that peeling of the deposit can be eliminated completely in this manner. This it can be supposed that the character of the internal stress has a profound effect on the adhesion, since peeling can be avoided by changing a tensile stress to a compressive stress.

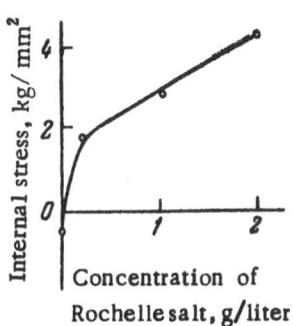

Fig. 81. The effect of the concentration of Rochelle salt on the internal stress in electrolytic copper deposits (Fedot'ev, Kruglova). Electrolyte containing 5 g/liter H_2SO_4; I_C = 1.0 amp/dm².

LITERATURE

1. P. Jacquet, Trans. Amer. Electrochem. Soc. 66 393 (1934).
2. N. P. Fedot'ev and E. G. Kruglova, Zhur. Priklad. Khim. 28, 275 (1955).

Chapter XIII

THE MECHANICAL PROPERTIES
OF THE ELECTROLYTIC DEPOSIT

The most important mechanical properties of the metallic deposit are its plasticity and elasticity. A solid body undergoes deformation when acted on mechanically, the term "deformation" being used here to designate any type of displacement of one part of the body with respect to another which results in an alteration of form and volume. Two types of deformation are to be distinguished: the static in which the applied force is in equilibrium with the body and the dynamic in which force and body are not in equilibrium and the body is set in motion. The deformation depends not only on the applied force but also on the type of motion acquired by the body. A deformation can be either elastic or plastic (inelastic). An elastic deformation disappears completely when the applied force is removed, while a plastic deformation persists even after removal of the deforming force (residual deformation). It is usually true that each body combines both elastic and plastic properties to a certain degree, one or the other being manifest to a greater extent, depending on the nature of the medium and the nature and structure of the metal. It is difficult to draw an exact line of demarcation between plastic and elastic deformation, since residual deformations tend to gradually disappear with the passage of time.

Anisotropy appears in the deformation of monocrystals, the slip being different in various crystallographic planes. The slip planes in the metallic monocrystal are usually cleavage planes, that is to say, planes of the highest atomic density and greatest interplanar spacing. A finely crystalline body usually deforms less readily than a body with coarsely crystalline structure, since the crystals are variously orientated in the latter and a slip in one crystal is inevitably retarded by neighboring crystals.

A basic factor in fixing the mechanical properties of the metal is the relative extension, a quantity which is proportional to the quotient of the deforming force, F, and surface area, S, or:

$$\frac{\Delta l}{l} = k \frac{F}{S}; \qquad \frac{F}{S} = E \frac{\Delta l}{l}. \tag{53}$$

In this equation, k is the coefficient of linear dilation, E is the Young modulus with the dimensions of stress, and l is the length of the body.

It is well known that the value of E is usually high and depends closely on the structure of the metal and the procedure followed in its preparation, especially if electrolytic methods are involved. The high value of E arises from the fact that this quantity is defined as the stress required for elastically deforming the body to twice its original length. Most bodies will rupture before a stress of this magnitude can be established. The minimum stress required for producing plastic deformation is designated as the elastic limit or flow limit. The maximum stress which a body cannot support without rupture is referred to as the tensile strength. The term fatigue is usually used to designate the breakdown of a metal under a periodic variation of stress, either with, or without, change of sign. It is observed that the tensile strength and fatigue limit are comparatively low in metals which contain nonmetallic occlusions distributed along grain boundaries and in cracks and fissures.

A very interesting and detailed, theoretical treatment of the effect of surface-active media on the deformation of metals is found in the text of Likhtman, Rebinder, and Karpenko [1].

THE HARDNESS OF THE ELECTROLYTIC DEPOSIT

It has been pointed out above that galvanic deposition can be used to profoundly alter the mechanical properties of a metal, especially hardness.

The data of the literature suggest that the hardness of an electrolytic deposit is always greater than the hardness of a sample of the same metal prepared by other methods [2]. Figure 82 gives a comparison of the hardness of various metals prepared in various ways. It is seen that the hardness of the electrolytic deposit is indeed higher than the hardness of the same metal prepared by other means, the difference being of considerable magnitude in certain cases.

The hardness of the electrolytic deposit is closely dependent on such factors as the composition of the electrolyte, the current density, and the temperature. It is not easy matter to compare hardness values when these have been obtained by different authors and are expressed in incompatible units. It is customary to measure the hardness in terms of the depth of penetration of a loaded sphere or diamond tetrahedron, since this will vary with the hardness of the test metal. A hardness value is obtained by dividing the load by the imprint area to give a quotient which is usually expressed in kilograms per square millimeter.

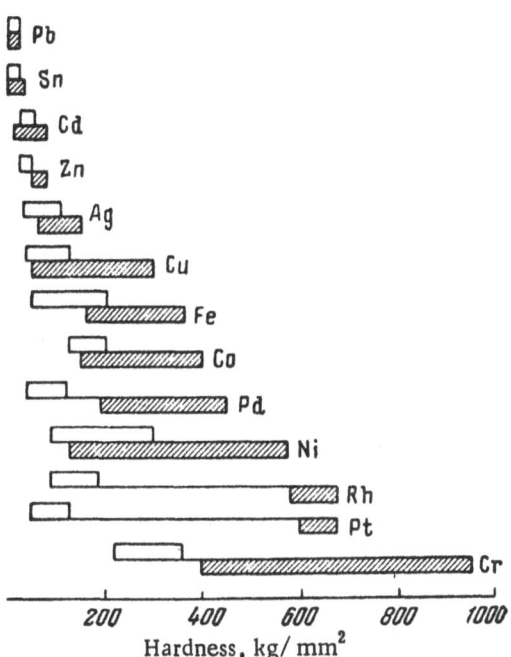

Fig. 82. A comparison of the hardness of various metals prepared in various ways (Fisher): □, metallurgical; ▨, galvanic.

THE EFFECT OF CURRENT DENSITY

A study of the effect of the current density on the hardness of chromium deposits has great practical interest since the hardness rises with the density in this case.

Experiment has proven that the hardness is not a simple function of the current density, the curve covering this relation showing both maxima and minima. As a rule, the hardness passes through a maximum as the density is altered. This maximum lies at 1000 kg/mm^2 and corresponds to a current density of 50-80 amp/dm^2 in the case of deposits obtained from standard electrolytes at 60° [3]. It is likely that alterations in deposit structure and quality account for the complexity of the relation between hardness and current density. There is a certain spread in the experimental data of various authors. It is probable that this situation arises from failure to assume uniformity of current distribution over the entire surface of the specimen, so that there is a sharp variation in hardness on passing from one point to another. There is an additional error from the marked alteration of surface relief with changing current density. All of these factors must be strictly controlled in hardness studies.

We have investigated the effect of current density on the hardness of nickel deposits[*] obtained from a standard electrolyte of pH 3.73 at room temperature (Fig. 83).

The figure shows that the hardness, current density curve (Curve 1) passes through a maximum at 360 kg/mm^2 and 3 amp/dm^2. The figure also shows the variation of hardness caused by heating the deposit in vacuum at 450-500° for one hour (Curve 2). Such heating reduces the hardness considerably, especially in the region of low current densities. This reduction is clearly due to the elimination of foreign substances (hydrogen, in particular) and the return of the lattice parameters to their normal values.

THE EFFECT OF TEMPERATURE

An elevation of temperature usually reduces the hardness of the electrolytic deposit. This is shown clearly by Fig. 84 which presents Rykova's data on the effect of temperature on the hardness of chromium deposits [3]. It is seen from the figure that the hardness falls from 700 to 400 kg/mm^2 when the electrolyte temperature is changed from 50 to 70°. Shreider [4] has obtained similar results and has pointed out that an increase of the temperature of chrome plating to 75° displaces the maximum on the hardness, current density curve toward densities in excess of 90 amp/dm^2.

The existence of a relation of this type has been confirmed by our own study of the electrodeposition of nickel (Fig. 85, Curve 1). Curve 2 of the figure shows the effect of electrolyte temperature on hardness after vacuum heating. It is seen that the most rapid fall-off in hardness occurs on raising the temperature from 20 to 40°, there being relatively little further reduction when the temperature is carried to 60°.

---

[*] Unless special mention is made to the contrary, hardness values where obtained on a PMT-3 apparatus and are expressed in kilograms per square millimeter.

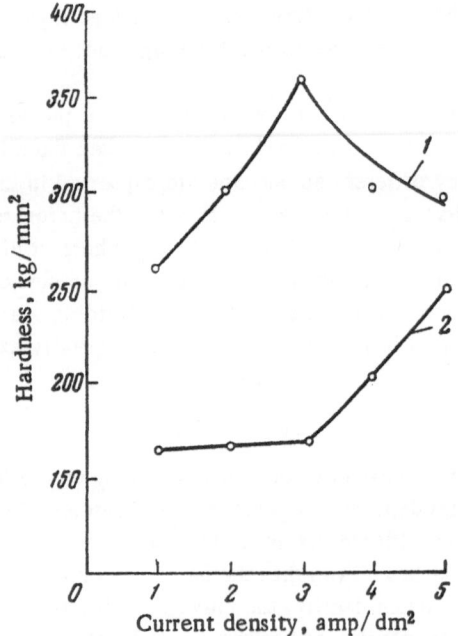

Fig. 83. The relation between current density and hardness of nickel deposits (Vagramyan, Petrova): Deposits obtained from an electrolyte containing: $NiSO_4 \cdot 7H_2O$, 266 g/liter; H_3BO_3, 30 g/liter; $Na_2SO_4 \cdot 7H_2O$, 24 g/liter; $MgSO_4 \cdot 10H_2O$, 20 g/liter at pH = 3.78 and t = 18-20°. 1) Prior to vacuum heating; 2) after vacuum heating.

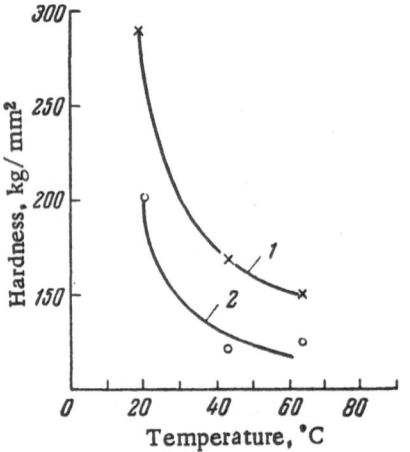

Fig. 85. The relation between the temperature of the electrolyte and the hardness of nickel deposits (Vagramyan, Petrova). Electrolyte composition, the same as in Fig. 83. 1) Prior to vacuum heating; 2) after vacuum heating.

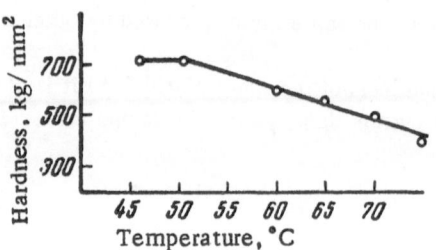

Fig. 84. The relation between temperature and hardness of electrolytic chromium deposits (Rykova). Chromium deposited from an electrolyte of standard composition at I_c = 80 amp/dm².

THE EFFECT OF THE ELECTROLYTE COMPOSITION

The hardness of the electrolytic deposit is clearly dependent on the composition of the electrolyte. Thus chromium deposits obtained from chromic anhydride solutions of different concentrations do not have the same hardness [3]. The data of Rykova on the effect of concentration on hardness are shown in Table 22.

The table makes it clear that an increase in the chromic oxide concentration decreases the deposit hardness considerably.

These figures are confirmed by the work of Shreider [4] which has shown that the hardness of a chromium deposit obtained at 65° from a solution containing 250 g/liter CrO_3 is 10-20% lower than the hardness of a deposit obtained from a less concentrated electrolyte (150 g/liter CrO_3) at the same temperature.

The composition of the solution is also of great significance in fixing the hardness of electrodeposits of nickel, cobalt, and iron. Thus, Fig. 86 shows the effect of the pH of the electrolyte on the hardness of iron deposits obtained from 1 N $FeSO_4$ at 25° and I_c = 2.0 amp/dm². Here, the hardness has been followed only through the 1.8-3.2 pH interval since it is not possible to prepare satisfactory iron deposits outside this range. Iron of maximum hardness, 320 kg/mm², is obtained from a ferrous sulfate solution of pH 2.8 (Fig. 86, Curve 1). The addition of 100 g/liter of aluminum sulfate to this solution lowers the hardness from 320 to 270 kg/mm² (Curve 2), while addition of boric acid at the same pH produces an abrupt rise in hardness to 430 kg/mm² (Curve 3). A comparison of hardness and occlusion can now be made on the basis of data on hydrogenation and hydroxide occlusion (Table 23).

The table brings out the fact that iron deposits obtained from 1 N $FeSO_4$ occlude essentially the same amount of hydrogen and hydroxides as iron deposits obtained from 1 N $FeSO_4$ containing added aluminum sulfate. The gas occlusion in a pure electrolyte is much less than the occlusion in a similar electrolyte containing boric acid. Thus, the presence of boric acid increases both the hardness of the iron deposit and the amount of gas occluded by it, although the relation between these factors is not a simple one.

The pH of the electrolyte also has a marked effect on the hardness. Thus, an increase in the pH of the electrolyte leads to a sharp rise in the hardness of nickel deposits obtained from an electrolyte containing: $NiSO_4 \cdot 7H_2O$, 166; H_3BO_3, 30; Na_2SO_4, 24; and, $MgSO_4$, 20 g/liter at 20° and a current density of 2.0 amp/dm². Here, the hard-

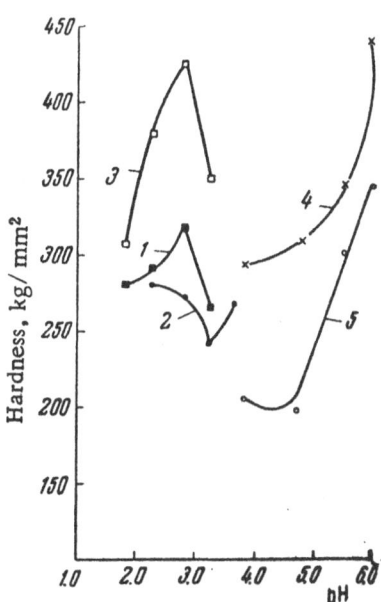

Fig. 86. The relation between the pH of the electrolyte and the hardness of nickel deposits (Vagramyan, Petrova): 1) Deposit obtained from 1 N $FeSO_4 \cdot 7H_2O$; 2) deposit obtained from a solution of this same type with 100 g/liter added $Al_2(SO_4)_3$; 3) deposit obtained from 1 N $FeSO_4 \cdot 7H_2O$ + 30 g/liter H_3BO_3 at I_c = 2.0 amp/dm² and t = 25°; 4) deposit obtained from the electrolyte of Fig. 83 at t = 18-20° and I_c = 2.0 amp/dm², prior to vacuum heating; 5) after heating.

TABLE 22. The Hardness of Electrolytic Chromium Obtained at 65° from Electrolytes Containing Various Concentrations of CrO_3

Current density, amp/dm²	Hardness (kg/mm²) at a CrO_3 concentration of:	
	250 g/liter	420 g/liter
25	595	318
50	642	366
80	954	597

ness reaches the rather high value of 400 kg/mm² at pH 6 (Fig. 86, Curve 4). Curve 5 covers the variation in hardness after vacuum heating. A comparison of curves makes it clear that the difference in hardness before and after heating is greater at lower pH values. This marked difference in hardness before and after heating at low pH probably arises from the fact that such hardness is due to hydrogenation and part of the hydrogen is driven off in heating. Hydroxides are occluded at high pH and remain as metallic oxides after heating so that no essential alteration in hardness results.

The effect of the electrolyte composition of the hardness of iron deposits has been investigated by Kudryavtsev and Yakovleva [5]. Results of a study of the effect of temperature and current density on the hardness of iron deposits obtained from electrolytes of various compositions are given in Table 24.

This table shows that the hardest iron deposits are those obtained from electrolytes containing potassium sulfate and oxalic acid. Kudryavtsev and Yakovleva have also proven that the addition of manganese sulfate has no essential effect on the hardness of iron.

The nature of the salt from which a metal is deposited has a great influence on the hardness. Thus, cobalt prepared from the 1 N sulfate does not have the same hardness as cobalt obtained from the 1 N chloride. A comparison of deposits of the best quality shows that cobalt prepared from cobalt sulfate is harder than cobalt obtained from cobalt chloride. This result is consistent with data on the occlusion of foreign substances.

Ioffe, Grubina, and Stroganov [6] have studied the hardness of nickel obtained from a solution containing: 1 N $NiSO_4$ + 0.7 g/liter Na_2SO_4 + 20 g/liter H_3BO_3 at 55° and I_c = 3.0 amp/dm², working over a wide interval of concentrations of added sodium chloride (Fig. 59, Curve 1). It is seen from the figure that the hardness of nickel rises sharply up to a NaCl concentration of 2 M and then remains practically constant as the concentration is increased further.

THE EFFECT OF ALLOYING METALS

It is well known that the physical properties of alloys are radically different from those of separate components. There is indication in the literature that the addition of even small amounts of foreign metals can profoundly alter such properties as hardness, elasticity, and tensile strength.

Thus, Morokhov and Atchi [7] have found that embrittlement of a nickel deposit can result from codeposition of lead that has passed into the electrolyte during dissolution of the bath lining. Interesting data on the hardness of

TABLE 23. The Effect of the Composition and pH of the Electrolyte on the Occlusion of Hydrogen and Hydroxides in Iron Deposits

pH	Gas occlusion cm^3/g					
	pure electrolyte		pure electrolyte + 100 g/liter $Al_2(SO_4)_3$		pure electrolyte + 30 g/liter H_3BO_3	
	hydrogen	hydroxides	hydrogen	hydroxides	hydrogen	hydroxides
1.80	4.30	1.34	—	—	7.9	—
2.25	3.50	1.50	3.76	1.50	7.0	—
2.80	3.80	1.54	3.64	1.56	17.5	—
3.20	2.28	1.50	3.42	1.60	19.5	—
3.50			2.56	2.46	20.0	11.5

TABLE 24. The Effect of the Electrolyte Composition on the Hardness of Iron

Electrolyte composition	Temperature, °C	Current density, amp/dm^2			
		1.0	2.0	3.0	4.0
		Hardness, kg/mm^2			
$FeSO_4 \cdot 7H_2O$ — 420 g/liter	20	290	340	—	—
pH 2.2-2.5	40	210	280	300	—
$FeSO_4 \cdot 7H_2O$ — 420 g/liter +	20	250	320	—	—
$Al_2(SO_4)_3 \cdot$ 100 g/liter; pH 3.5	40	190	300	320	340
$FeSO_4 \cdot 7H_2O$ — 420 g/liter + K_2SO_4	20	550	600	615	—
150 g/liter; pH 2.2-2.5	40	300	350	390	—
$FeSO_4 \cdot 7H_2O$ — 420 g/liter +	20	500	515	605	635
$H_2C_2O_4$ — 4 g/liter; pH 2.2-2.5	40	250	295	360	—
$FeSO_4 \cdot 7H_2O$ — 420 g/liter + K_2SO_4 — 150 g/liter; $H_2C_2O_4$ — 4 g/liter	20	520	532	615	640
pH 2.2-2.5	40	300	340	450	570

gold deposits have been obtained by Atanasyants, Kudryavtsev, and Karataev [8]. Here, it was shown that the presence of some 2% nickel would increase the hardness of gold deposits from 100 to 160-190 kg/mm^2, and the same time, give a 12-fold increase in the wear resistance if deposition was at elevated temperature or a 4-fold increase if deposition was at room temperature. For the preparation of hard gold deposits these authors recommend the use of an electrolyte containing: gold, 1.5; nickel, 3.5; free potassium cyanide, 15; and, potassium carbonate, 35 working at I_c = 0.2 amp/dm^2 and 20-60°.

TABLE 25. The Effect of Mn, Cu, and Fe Sulfates on the Hardness of Nickel Deposits Obtained from an Electrolyte Containing: $NiSO_4$, 210; H_3BO_3, 30; and, KCl, 10 g/liter working at I_c 2.0 amp/dm^2, 20°, and pH, 3.78

Additive, g/liter	Hardness, kg/mm^2	
	prior to vacuum heating	after vacuum heating
Mn-0.3	392	382
Cu-0.3	Loose deposits	Loose deposits
Fe-0.3	310.5	240
Without additive	300.0	210

Our own data indicate that an alteration in the hardness of nickel deposits can be brought about by the addition of 0.3 g/liter of manganese and copper sulfates (Table 25).

This table makes it clear that the hardness of electrolytic nickel deposits is practically unaffected by addition of iron and increased rather considerably by the presence of manganese.

THE EFFECT OF SURFACE-ACTIVE SUBSTANCES

The hardness of the deposit can usually be increased by the addition of surface-active substances at definite concentrations, a fact illustrated by our own data of Table 26.

This table brings out the fact that the hardness of electrolytic deposits of copper and nickel is increased considerably by the addition of sodium naphthalenedisulfonate and thiourea to the electrolyte, the magnitude of the effect depending on the concentration of the additive. Curve 1 of Fig. 87 covers the relation between thiourea concentration and the hardness of copper deposits and shows the greatest increase in hardness to occur with 0.005 g/liter

TABLE 26. The Effect of Surface Active Substances on the Hardness of Deposits of Copper and Nickel

Composition of electrolyte and conditions of electrolysis	Hardness, kg/mm^2	Composition of Electrolyte and conditions of electrolysis	Hardness, kg/mm^2
$NiSO_4 \cdot 7H_2O$— 210, H_3BO_3—30 and KCl— 10 g/liter; pH 3.78; 20°; I_c = 2-3 amp/dm^2	300-337	$CuSO_4 \cdot 5H_2O$— 250 and H_2SO_4—50 g/liter, 20°; I_c = 3.0 amp/dm^2	100
		The same electrolyte with 0.5 g/liter added sodium naphthalenedisulfonate	160
The same electrolyte with 5 g/liter added sodium naphthalenedisulfonate	611	The same electrolyte with 0.005 g/liter added thiourea + 0.5 g/liter added sodium naphthalene	230-260
The same electrolyte with 0.15 g/liter added thiourea	877	The same electrolyte with 0.015 g/liter added thiourea	248
		The same electrolyte with 20 g/liter	146
		The same electrolyte with 20 g/liter + 20 ml/liter glycerin	160

of this additive. Further addition of thiourea is practically without effect on the copper hardness. Khonikevich and Fedot'ev [9] and Fedot'ev and Pozin [10] have obtained similar results with Rochelle salt. Here, it was shown that the presence of 0.0025 g/liter of the additive would increase the hardness of the copper deposit from 152 to 263 kg/mm^2.

THE EFFECT OF ALTERNATING AND REVERSING CURRENTS

Deposition under an alternating current makes it possible to profoundly alter the structure and hardness of the deposit. Rykova [3] has studied the effect of a periodic current interruption on the structure and hardness of chromium deposits and has shown that high-quality plates can always be obtained by laying down laminar deposits in which the separate layers are each approximately 3 μ thick. Such deposits are homogeneous, smooth, compact, and brilliant. Their hardness is closely dependent on the current density, the maximum hardness at I_c = 100 amp/dm^2 being obtained with a deposition period of 10 min and an interruption period of 1 min.

The application of a reversing current proves to be very beneficial in improving both quality and hardness in certain electrodepositions.

Thus, Bakhvalov and Rumyantsev [11] have shown that copper and zinc deposits obtained from acidic electrolytes under a current of reversing polarity are more finely grained in crystalline structure, harder, and less porous than deposits obtained with a direct current.

Skirstymonskaya [12] has studied the electrodeposition of copper and zinc with a superimposed alternating current, finding little alteration in the copper structure at low alternating densities and a coarsening of the copper crystals with reduction in hardness when the density is high. Khonikevich and Fedot'ev have obtained very similar results in a study of the deposition of copper under superimposed alternating and direct currents [9]. These authors have shown that the hardness of a deposit obtained with a $T_a : T_c$ = 1 : 10 current reversal is always reduced by adding thiourea (Fig. 87, Curves 2 and 3), Rochelle salt (Fig. 88), or gelatine (Fig. 89) to the electrolyte.

It follows from these figures that a reversing current can variously affect the hardness. It is natural to expect an increase in hardness when the reversing current increases deposit occlusion, and a diminution in hardness in the opposite case. The structure is affected by the number of adsorbed particles and there should be a relation between deposit hardness and crystal dimensions. It can be generally assumed that the hardness will always be increased by carrying out deposition under conditions favoring a reduction in grain size. Thus, Macnaughtan and Hothersall [13] have claimed that the hardness of nickel deposits rises as the crystal structure becomes finer and finer. This same conclusion is supported by Phanhauser [14] (Table 27).

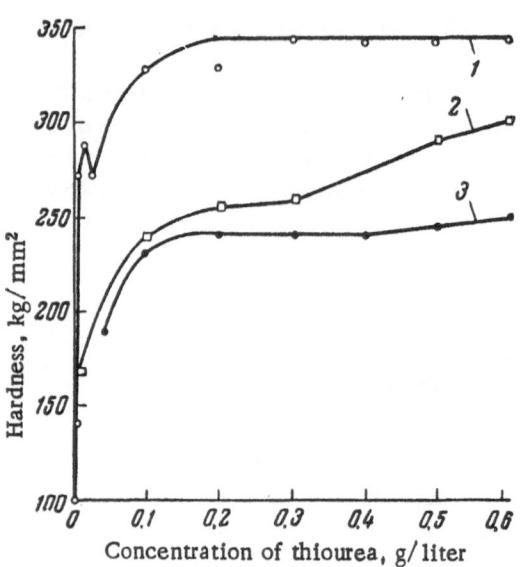

Fig. 87. The relation between the concentration of thiourea and the hardness of copper deposits: 1) Deposit obtained from a standard electrolyte; t = 18-20°; I_c = amp/dm^2 (Tsareva, Solokhina, Kudryatseva, Vagramyan); 2) deposit obtained from an electrolyte of this same composition at I_c = 2.0 amp/dm^2 (Khonikevich, Fedot'ev); 3) hardness for a 1 : 10 current reversal (Khonikevich, Fedot'ev).

THE EFFECT OF ULTRASONIC WAVES

Study has shown that ultrasonic waves have a marked effect on the mechanical properties of the electrolytic deposit. Thus, it has been shown that the hardness of nickel can be raised from 215 kg/cm^2 (normal value) to 310 kg/cm^2 by carrying out electrodeposition in a 16-kc ultrasonic field [15]. Electrodeposition in an ultrasonic field also results in a marked increase in the hardness of chromium and copper. Figure 90 presents the data of Miller and Kuss [15] on the hardness, current density relation in chromium deposits obtained under ordinary conditions, and in an ultrasonic field. These same authors have also reported high tensile strength in electrolytic deposits obtained in an ultrasonic field. Such ultrasonic effects can be considered as resulting from intensive agitation of the electrolyte which accelerates delivery of foreign particles to the electrode and thereby promotes occlusion in the deposit.

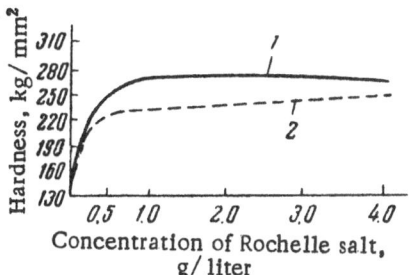

Fig. 88. The relation between the concentration of Rochelle salt and the hardness of electrolytic copper deposits (Khonikevich, Fedot'ev): 1) Without superposition of an alternating current; 2) with a 1 : 10 current reversal.

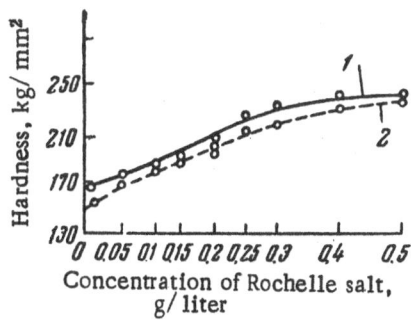

Fig. 89. The relation between the concentration of gelatine and the hardness of electrolytic copper deposits (Khonikevich, Fedot'ev): 1) Without the application of an alternating current; 2) with a 1 : 10 current reversal.

TABLE 27. The Effect of Structure on the Hardness of Electrolytic Deposits

Metallic deposits	Hardness, kg/mm^2	Scratch hardness, relative values
Coarsely crystalline copper from acid bath	70-100	50
Finely crystalline copper from cyanide bath	135-210	59
Coarsely crystalline nickel from hot bath	212-240	39
Finely crystalline nickel from cold bath	390-440	72
Dull chromium from cold bath	600-650	150
Shiny chromium from bath heated to 45°	750-900	250
Hard, shiny chromium from bath heated to 55°	900-1100	–

THE CAUSE OF HARDNESS

The experimental results which have been cited above show that the hardness of the electrolytic deposit is determined by the occlusion of hydrogen, metallic oxides, surface-active substances, and other foreign materials. The occlusion of one substance or another will be favored, depending on the metal and the conditions of electrolysis.

Foerster and Lee [16] consider the reduction of hardness with rising temperature to be due to a diminution in the hydrogen occlusion. This same point of view is maintained by Revyakin [17] who has drawn attention to the fact that the hardness of deposits of iron group metals is determined by the hydrogen occlusion. According to Arkharov and Nemnonov [18], the exceptional hardness of electrolytic chromium deposits is due to crystal lattice distortions arising from internal stress and hydrogen atom penetration.

Mamontov and Petrov [19] claim distortion of the crystal lattice is the principle cause of the enhanced hardness of electrolytic deposits. The studies of these authors have disclosed certain differences in the effect of heating on deposit hardness. Thus, a high initial hardness is maintained unaltered on heating to 300° and begins to diminish only when the temperature is raised still further, whereas, a low initial hardness increases somewhat on heating to 300°, passes through a maximum, and then falls at temperatures in excess of 500°. There are no maxima on the lattice parameter, temperature and hardness, temperature curves when surface-active substances are present and the initial hardness is rather high.

The data of Mamontov and Petrov on the alteration of the crystal lattice parameters with the temperature of heating are presented in Table 28.

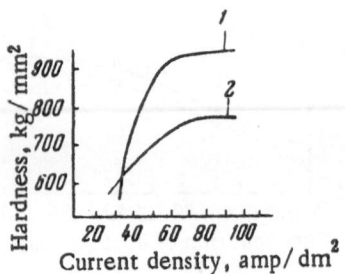

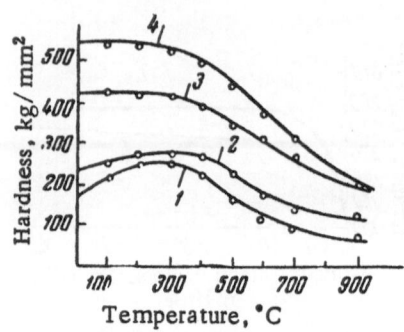

Fig. 90. The relation between the current density and the hardness of electrolytic chromium deposits (Miller, Kuss): CrO_3, 250 g/liter; H_2SO_4, 2.5 g/liter; t = 50°: 1) In an ultrasonic field; 2) under ordinary conditions.

Fig. 91. The relation between the temperature and the hardness of iron deposits. (Mamontov, Petrov: 1) electrolyte containing: $FeCl_2 \cdot 4H_2O$, 500 g/liter; NaCl, 100 g/liter; HCl, 2.2 g/liter; I_C = 18 amp/dm²; t = 90°; 2) electrolyte containing: $FeCl_2 \cdot 4H_2O$, 500 g/liter; NaCl, 100 g/liter; HCl, 2.0 g/liter; glycerin, 100-150 g/liter; I_C = 18 amp/dm²; t = 80°; 3) electrolyte containing: $FeCl_2 \cdot 4H_2O$, 500 g/liter; NaCl, 100 g/liter; glycerin, 100 g/liter; sugar, 40 g/liter; HCl, 1.6 g/liter; I_C = 18 amp/dm²; t = 90°.

This table shows that the parameters remain constant up to 300° and diminish with further rise in temperature to 700°. These authors also note that the introduction of surface-active substances increases the recrystallization temperature of the deposit markedly.

It has been brought out in the review of metal hydrogenation and adsorption of surface-active substances that foreign substances can exist in the electrolytic deposit in various states (ions, molecules, partial dissociation) and be

TABLE 28. The Temperature Variation of the Lattice Parameters

Composition of electrolyte, g/liter				Conditions of electrolyte		Temperature of heating deposit	Lattice constant, A
$FeCl_2 \cdot 4H_2O$	NaCl	glycerin	HCl	I_C, amp/dm²	t, °C		
500	100	—	2.2	18	90	—	2.8608
500	100	—	2.2	18	90	300	2.8610
500	100	—	2.2	18	90	500	2.8593
500	100	—	2.2	18	90	700	2.8592
500	100	—	2.2	18	90	900	2.8592
500	100	120	2.0	20	90	—	2.8606
500	100	120	2.0	20	90	300	2.8606
500	100	120	2.0	20	90	500	2.8594
500	100	120	2.0	20	90	700	2.8592
500	100	120	2.0	20	90	900	2.8592

variously distributed (in pores or cracks, on grain boundaries, in the lattice, etc.). It is clear that the effect of a given amount of foreign material on the hardness will vary with its state and distribution in the deposit. This explains the fact that the hardness increases with increasing occlusion of various types, even though there is no strict relation between these factors.

LITERATURE

1. V. I. Likhtman, P. A. Rebinder, and G. V. Karpenko, The Effect of Surface-Active Media on the Deformation of Metals [in Russian] (Acad. Sci. USSR Press, Moscow, 1954).
2. H. Fischer, Elektrolytische Abscheidung und Elektrokristallisation von Metallen (Berlin, 1954).
3. A. V. Rykova, Studies on the Corrosion of Metals under Stress [in Russian] (Mashgiz, Moscow, 1953).
4. A. V. Shreider, Collected Papers on the Theory and Practice of Electrolytic Chrome Plating [in Russian] (Acad. Sci. USSR Press, Moscow, 1957) p. 77.
5. N. T. Kudryavtsev and L. A. Yakovleva, Tr. Gos. NII GVF (Moscow, 1957) p. 26.
6. V. S. Ioffe, Grubina, and Stroganov, Otchet GIPKh (1941).
7. M. I. Morokhov and A. P. Atchi, Korroziya i Bor'ba s Nei 5-6, 10 (1940).
8. A. G. Atanasyants, N. T. Kudryavtsev, and V. M. Karataev, Zhur. Priklad. Khim. $\underline{30}$, 876 (1957).
9. A. A. Khonikevich and N. P. Fedot'ev, Tr. Leningr. Tekhnol. In-ta. im. Lensoveta, 40, (Goskhimizdat, Leningrad, 1957).
10. N. P. Fedot'ev and Yu. M. Pozin, Zhur. Priklad. Khim. $\underline{31}$, 424 (1958).
11. G. T. Bakhvalov and N. V. Rumyantsev, Electroplating of Metals with a Reversing Current, Copy of a lecture in Mosk. Dom. N.-T. Prop. im. F. E. Dzerzhinskogo [in Russian] (1957).
12. V. I. Skirstymonskaya, Zhur. Fiz. Khim. $\underline{10}$, 617 (1937).
13. D. J. Macnaughtan and A. W. Hothersall, Trans. Faraday Soc. $\underline{24}$, 387 (1928); $\underline{31}$, 1168 (1935).
14. F. Phanhauser, Galvanotechnik (Leipzig, 1949) pt. I, p. 105.
15. F. Miller and H. Kuss, Helv. chim. acta $\underline{33}$, 217 (1950).
16. F. Foerster, Elektrochemie wässeriger Lösungen (Leipzig, 1922).
17. V. P. Revyakin, Izvest. Vyssh. Uch. Zav., Fizika 1, 132 (1958).
18. V. I. Arkharov and S. A. Nemnonov, Zhur. Tekh. Fiz. $\underline{8}$, 1089 (1938).
19. E. A. Mamontov and Yu. N. Petrov, Leningr. Gos. Ped. In-t. im. A. I. Gertsena, Uch. Zap. Kaf. Fiz. i Mat. 141 (1958).

Chapter XIV

THE RELATION BETWEEN INTERNAL STRESS, HARDNESS, AND ELASTICITY

Hume-Rothery and Wyllie [1] consider that there must be a relation between the structural orientation in the electrolytic deposit and the internal stress, hardness, and elasticity.

The theory advanced by these authors claims that the deposit structure is fixed by the working conditions and that this, in turn, completely determines the properties. Figure 92 presents experimental data pointing to the existence of a relation between structure, internal stress, hardness, and elasticity in chromium deposits obtained from a standard chromic acid electrolyte containing H_2SO_4 in a 100 : 1 ratio, working at various temperatures with $I_c = 110$ amp/dm².

The temperature range covered by this figure clearly divides into four sub-intervals. The first of these (I) covers the 25-50° range and corresponds to disoriented deposit structures. Traces of a (111) type orientation are observed on the boundary between regions I and II. Region II covers the 50-62.5° range and corresponds to an increase in the number of oriented crystals, maximum orientation being observed over the hatched interval ranging from 62.5 to 76.5°. The number of oriented crystals again diminishes with increasing temperature over region III, which covers the 67.5-80° range. Crystal orientation disappears completely on the boundary between regions III and IV, while disoriented deposits are obtained in IV itself. These structural changes are accompanied by corresponding alterations in the internal stress, hardness, and elasticity of the deposit. The internal stress increases sharply with rising temperature over the first interval and reaches a maximum of 4650 kg/cm² at the I-II boundary. Continued increase in temperature eventually gives deposits of maximum orientation and minimum internal stress. The internal stress again rises with further increase in temperature, reaches a second maximum one-third as high as the first, and then falls over region IV. The curves show the elasticity of the deposit to be at a minimum when the orientation is at a maximum. The hardness of the chromium deposit is at a maximum when the orientation is greatest, and falls off over regions III and IV. These data point to the existence of a complex interrelation between orientation, internal stress, hardness, and elasticity.

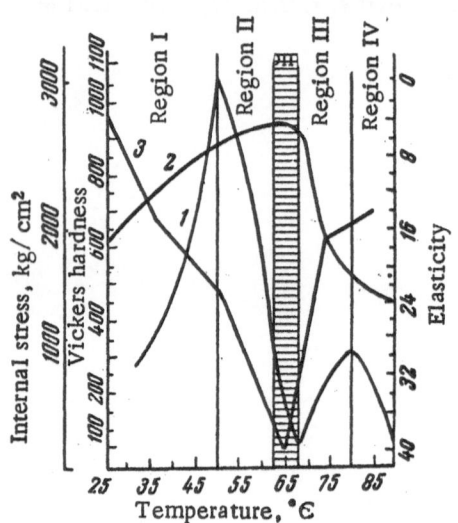

Fig. 92. The relation between internal stress, hardness, and elasticity (Hume-Rottery, Willie): 1) Internal stress; 2) hardness; 3) elasticity.

LITERATURE

1.   W. Hume-Rothery and M. K. Wyllie, Proc. Roy. Soc. A, <u>182</u>, 131 (1943).

Chapter XV

THE EFFECT OF ELECTROCHEMICAL TREATMENT ON BRITTLENESS

Brittleness is a very undesirable quality of an electrolytic deposit which is closely tied up with the preliminary chemical and electrochemical treatment of the base and the conditions of subsequent deposition of the plate. Excessive brittleness is the cause of rapid breakdown of machine parts in operation.

Brittleness can result from the occlusion of hydrogen, foreign metals, surface-active substances, and other materials.

Brittleness is generally measured by the amount of bending or twisting required for cracking the deposit.

Tomashov, Tyukina, and Blinchevskii [1] have constructed an apparatus for measuring brittleness in which a test sheet is bent by the upward movement of a special screw-operated device. Measurement is made of the bend angle required for cracking the sheet. This angle characterizes the brittleness of the deposit, high values of the one quantity corresponding to low values of the other. Only relative values of the brittleness can be obtained by this method and the results are affected by such extraneous factors as the nature of the base and the thickness of the test sheet.

Another method of determining the brittleness involves observation of the number of twists required for breaking the specimen. The brittleness is, in this case, characterized by the torsional resistance and is expressed in percentages by the equation:

$$B_x = \frac{100 \cdot x}{x_0}, \qquad (54)$$

x being the number of twists required for breaking the hydrogenated, and x_0, the number required for breaking the unhydrogenated, specimen.

THE EFFECT OF PRELIMINARY TREATMENT ON THE BRITTLENESS OF A METAL

There is a rather extensive literature on hydrogen embrittlement during preliminary treatment of the metallic surface. Special mention should be made of the work of Smyalovskii and his co-workers [2] on the effect of various factors on this type of embrittlement.

THE EFFECT OF CURRENT DENSITY

Figure 93 illustrates the effect of current density on the hydrogen embrittlement of iron polarized in $1 \, N \, H_2SO_4$. The figure indicates that the torsional resistance falls off sharply with increasing current density, reaches a value of approximately 60% at 15 amp/dm^2, and then remains practically constant.

THE EFFECT OF THE COMPOSITION OF THE SOLUTION

Smyalovskii and Shklyarskaya-Smyalovskaya [2] have also studied the effect of solution composition on the torsional resistance of iron, and have shown that the brittleness of this metal can be altered considerably by various additives. Thus, the contaminants which are present in pure sulfuric acid are sufficient to reduce the torsional resistance by some 10% in this case (Curve 2). The presence of small quantities of arsenic in sulfuric acid solutions can reduce the torsional resistance to one-sixth of its original value (Curve 3).

Study has shown that antimony, selenium, and tellurium have the same effect on brittleness as arsenic. Smyalovskii and Shklyarskaya-Smyalovskaya consider that these additives poison the active-surface centers for recombination of hydrogen atoms and favor the occlusion of these atoms in the metal.

On the other hand, dibenzylsulfoxide will counteract the effect of arsenic (Fig. 94). The figure makes it clear that this additive will increase the torsional resistance both in pure sulfuric acid and in sulfuric acid containing added arsenic. It is to be noted that the effect of this compound is closely dependent on the pH of the solution, and that it has no influence on the brittleness in neutral and alkaline media.

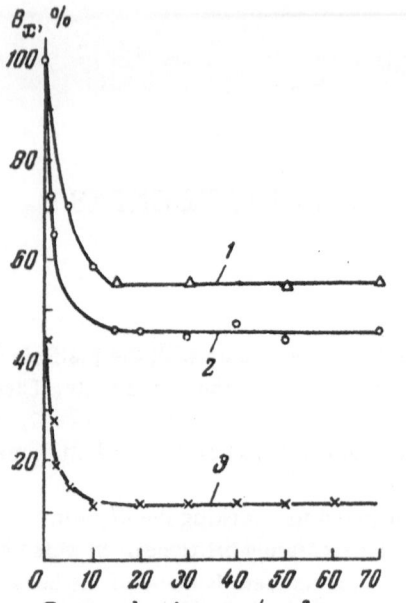

Fig. 93. The effect of current density on the hydrogen embrittlement of iron (Smyalovskii, Shklyarskaya-Smyalovskaya): 1) In 1 N H_2SO_4 (high purity); 2) in 1 N H_2SO_4 (pure); 3) in 1 N H_2SO_4 + As.

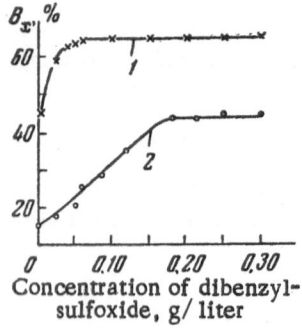

Fig. 94. The effect of dibenzylsulfoxide on the brittleness of iron in the presence of arsenic (Smyalovskii, Shklyarskaya-Smyalovskaya): 1) 1 N H_2SO_4; 2) 1 N H_2SO_4 + As.

Kudryavtsev and Moroz [3] have studied the relation between hydrogen embrittlement in a steel, the acid used for its etching and pickling, and the additives which this acid contains. Their work has shown that the brittleness of steel is scarcely affected by pickling in 5% HCl and etching at 18° in a 10% HCl solution containing six drops per liter of "KS."** The brittleness is altered very little by etching in a similar solution which is free of "KS." Even a few minutes etching in 20% H_2SO_4 sharply increases the brittleness, but this effect can be reduced somewhat by the addition of "ChM"*** to the solution. These experiments show that etching in HCl containing "KS" is a satisfactory means of eliminating brittleness.

Figel'man and Shreider [4] have studied the effect of various fa factors on the hydrogen embrittlement of steel, determining the brittleness by the bending method and expressing the results in percentages referred to a nonhydrogenated specimen. This work covers the effect of naphthalenedisulfonic and nitric acids, dextrine, formaldehyde, citric acid, chromic anhydride, oxalic acid, ammonium thiocyanate, phenol, aminophenol, "KS," "ChM," and potassium permanganate, on the hydrogen embrittlement of a steel cathodically polarized in a 10% H_2SO_4 solution for 600 sec at 20° and a current density of 2.0 amp/dm^2. It was found that almost all of these additives would reduce the brittleness of annealed steel, the effect of chromic anhydride being specially marked. The authors claim that the reduction of brittleness by chromic anhydride results from an oxidation of atomic hydrogen by this compound. Addition of ammonium thiocyanate increases the brittleness because of catalytic action by the sulfides which are formed in the breakdown of the thiocyanate ion.

Study has shown that the hydrogen embrittlement of steel cathodically polarized in 10% NaOH solution is less than that observed in acid solutions. The presence of oxidizing additives does not reduce the brittleness in this case.

Most of the additives studied by Figel'man and Shreider increase the brittleness of annealed and chilled steels. Sodium nitrate, sodium thiosulfate, and potassium thiocyanate have a pronounced effect on the hydrogen embrittlement of chilled steel. Sodium nitrate, potassium chromate, potassium thiocyanate, sodium thiosulfate, and sodium sulfate are most effective in increasing the brittleness of annealed steel. Figel'man and Shreider also found that the addition of 1% NaCN to an alkaline solution would produce a marked increase in the brittleness of annealed and quenched steels.

THE EFFECT OF THE TIME OF POLARIZATION

Figure 95 shows that the torsional resistance is sharply reduced by even a very brief polarization of 1-3 min, the reduction becoming sharper and more extensive as the hydrogen ion concentration of the solution is increased. Figure 98 presents data on the effect of arsenic on the brittleness of iron.

---

*Sulfonated blood serum.

**A mixture of a foaming agent and an etching regulator.

A comparison of Figs. 95 and 96 shows that arsenic increases the hydrogen embrittlement of iron in every case. The situation in regard to the hydrogenation of nickel is somewhat different (Fig. 97). Here, the change of brittleness with time of cathodic polarization is less marked than in the case of iron and a longer time is, therefore, required for reaching the maximum. Thus, a 45% reduction of the torsional resistance of iron is reached in a 10 min cathodic polarization, while the same reduction of resistance in nickel requires a 120 min polarization.

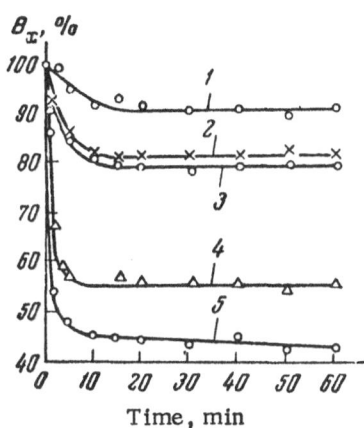

Fig. 95. The effect of the time of polarization on the hydrogen embrittlement of iron in various electrolytes (Smyalovskii, Shklyarskaya-Smyalovskaya): 1) 1 N NaOH; 2) 1 N N₂SO₄; 3) 0.01 N H₂SO₄; 4) 0.1 N H₂SO₄; 5) 1 N H₂SO₄.

It should be pointed out that hydrogen adsorption studies [5] have shown the effect of catalytic poisons on brittleness to be much less pronounced in nickel than in iron.

THE EFFECT OF THE TIME OF STORAGE

Smyalovskii and Shklyarskaya-Smyalovskaya [2] have studied the effect of time of storage at room temperature on the properties of hydrogenated iron and nickel. This study has proven that nickel is almost fully regenerated after 48 hr, while iron fails to regenerate appreciably even over extended periods (Fig. 98). The data indicate that the elimination of hydrogen effects in the presence of arsenic, a hydrogenation promoter, is more rapid in nickel than in iron.

It is to be noted that the treatment given to the metallic surface prior to plating also has a pronounced effect on the fatigue limit. Thus, etching in HCl will, on the average, reduce this limit by 18% [6], while a 10% reduction results from etching in 5% H₂SO₄ at 66° with subsequent elimination of hydrogen embrittlement by heating to 100° [7]. These observations are in qualitative agreement with the data on brittleness. The fatigue limit falls with an increase in the time and temperature of etching [8].

THE EFFECT OF ELECTROLYTIC DEPOSIT ON BRITTLENESS

It frequently happens that the brittleness of a metal is increased by the electrolytic deposition of a plate on it. Study has shown that there is a parallelism between brittleness and internal stress, so that any factor which affects the one will affect the other, as well. Figure 99 presents data on the effect of surface-active substances on the brittleness and internal stress of copper deposits obtained in the presence of thiourea. This figure shows that the brittleness increases with rising thiourea concentration, the alteration being quite pronounced up to a concentration of 0.005 g/liter. This increase in brittleness is paralleled by a rise in internal stress.

Experimental studies have shown [4] that deposition embrittlement of the base metal is completed in the first few minutes of electroplating and remains essentially constant when electrolysis is carried beyond this point. The time required for maximum reduction in this embrittlement depends on the nature of the base and the conditions of deposition. Thus, the brittleness of steel, zinc plates in a cyanide bath, is considerably higher than that of steel plates in a sulfuric acid bath.

Kudryavtsev and Moroz [3] have also studied the effect of various types of electrolytic zinc plating. Their work has shown that the electrolyte composition, conditions of electrolysis, and preliminary thermal treatment are all significant factors here. In particular, the brittleness, current density relation varies with the type of preliminary thermal treatment (Fig. 100).

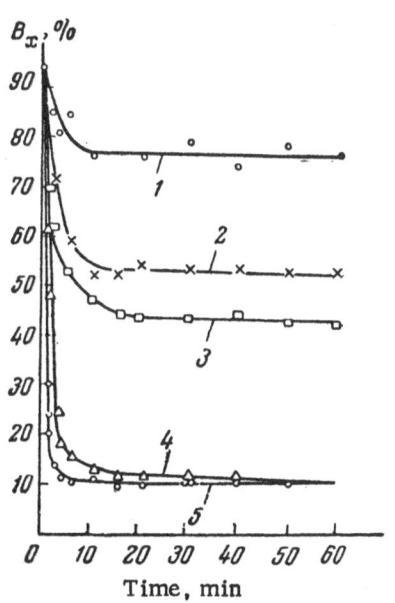

Fig. 96. The effect of arsenic on the brittleness of iron (Smyalovskii, Shklyarskaya-Smyalovskaya): 1) 1 N Na₂SO₄; 2) 1 N Na₂SO₄; 3) 0.01 N H₂SO₄; 4) 0.1 N H₂SO₄; 5) 1 N H₂SO₄.

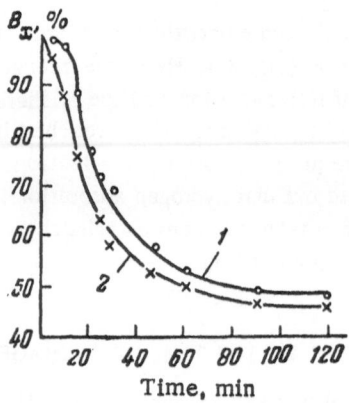

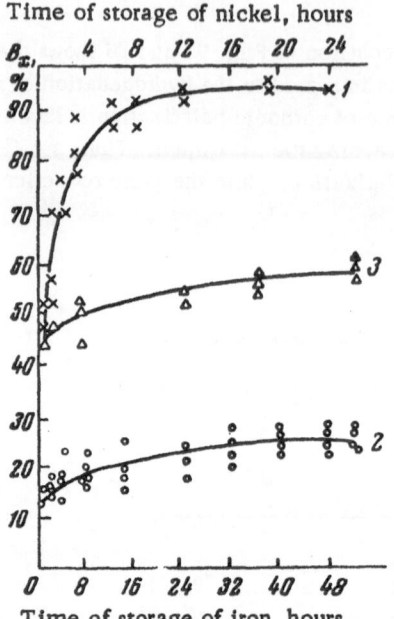

Fig. 97. The effect of the time of cathodic polarization on the brittleness of nickel (Smyalovskii, Shklyarskaya-Smyalovskaya):
1) 1 N H_2SO_4; 2) 1 N H_2SO_4 + As.

Fig. 98. The relation between hydrogen embrittlement in iron and nickel and the time of storage (Smyalovskii, Shklyarskaya-Smyalovskaya): 1) Nickel with added arsenic; 2) iron with added arsenic; 3) pure iron.

The figure makes it clear that an increase in current density from 0.5 to 4 amp/dm^2 will not essentially alter the plastic properties of 30KhGSA* steels tempered at various temperatures, but will considerably reduce the brittleness of high-carbon steel. The authors point out that such an increase in current density diminishes the time required for forming a deposit of given depth, thus reducing the degree of hydrogenation and the hydrogen embrittlement of a high-carbon steel.

Studies of the influence of the electrolyte compositions have shown a variation in the increase of hardness obtained with zinc plates prepared from cyanide, ammoniacal, and sulfate baths, both with and without additives. The best results are obtained when zinc plating is carried out from a fluoborate electrolyte containing: $Zn(BF_4)_2$, 250; $(NH_4)BF_4$, 25; and licorice, 2 g/liter at I_C = 5 amp/dm^2 and 20°.

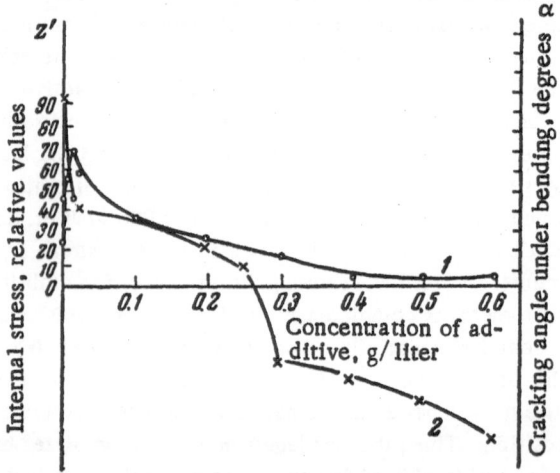

Fig. 99. The effect of the thiourea concentration on the brittleness (1) and internal stress (2) of copper deposits (Tsareva, Solokhina, Kudryatsev, Vagramyan).

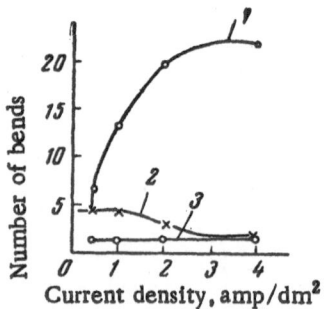

Fig. 100. The effect of the current density on the hydrogen embrittlement of steels which have been subjected to various types of thermal treatment (Kudryavtsev, Moroz): 1) High-carbon steel; 2) 30KhGSA steel, quenched at 880°, tempered at 500°; 3) 30 KhGSA steel, quenched at 880°, tempered at 200°

---

*30KhGSA steel has the composition: C—0.30; Mn—1.00; Si—1.00; Cr—1.00; Ni—0.40; S—0.03; P—0.035.

Many types of steel show no increase in brittleness in this electrolyte.

Studies on the brittleness of zinc plated steels have proven that reversing currents can be used at high densities to obtain quality deposits that have lower internal stress than deposits prepared under standard conditions. Kudryavtsev and Moroz consider that the principal factor accounting for changes in brittleness during electrolysis is the alteration in the rate of evolution of hydrogen resulting from variation in conditions. An increase in the amount of evolved hydrogen will increase the amount of this gas diffusing into the deposit and thus enhance the brittleness.

LITERATURE

1. N. D. Tomashov, M. N. Tyukina, and G. K. Blinchevskii, Studies on the Corrosion of Metals [in Russian] (Acad. Sci. USSR Press, Moscow, 1951).

2. M. Smyalovskii and Z. Zhklyarskaya-Smyalovskaya, Prace Konferencji Elektrochemicznej. Warszawa, Polska Akademia Nauk, 1957, 301.

3. N. T. Kudryavtsev and I. I. Moroz, The Elimination of Hydrogen Embrittlement during Electrolytic Zinc Plating [in Russian] (Mock. Dom. N.-T. Prop. im. F. E. Dzerzhinskii, 1957).

4. M. A. Figel'man, Hydrogen Embrittlement in the Cathodic Treatment of Steels and Methods of Combating It. Scientific-Technical and Pilot Experiments (Acad. Sci. USSR, Filial VINITI, Tema 13, No. M-57-259/14, Moscow, 1957); M. A. Figel'man and A. V. Shreider, Zavodskaya Lab. 22, 586 (1956); Zhur. Priklad. Khim. 31, 1184 (1958).

5. N. N. Kavtaradze, Zhur. Fiz. Khim. 32, 1214 (1958).

6. W. H. Swanger and J. J. Gongh, J. Iron and Steel Inst. 135, 315 (1937).

7. J. H. Fryl and R. Z. Manteufell, Automobiltechn. Zs. 46, 304 (1943).

8. G. L. Kehl and C. M. Offenhauer, Trans. Amer. Soc. Metals 28, 238 (1940).

Chapter XVI

THE EFFECT OF THE ELECTROLYTIC DEPOSIT
ON FATIGUE STRENGTH AND DURABILITY

It has already been pointed out that the galvanic method is widely used for plating machine parts which are to operate under heavy load or be subjected to severe corrosion. This method is also applied in cases where high durability is demanded of the part in question.

A number of investigations have shown that the fatigue strength is reduced sharply by a galvanic plate under tensile stress. Thus, considerable interest attaches to a study of the effect of various technological electrochemical factors on the fatigue limit.

The most commonly used platings are prepared from chromium, nickel, and copper. Durable chromium plates are frequently applied to highly loaded machine parts which are repeatedly acted upon by force of varying magnitude and sign.

Work in this field was initiated by the detailed studies of Barklie and Davies [1] on the effect of nickel, zinc, and lead plates on the fatigue strength of metals under periodically varying loads. Here, it was shown that the internal stress would sharply reduce the fatigue strength of a steel machine part which had been nickel plates from a solution containing: $NiSO_4 \cdot 7H_2O$, 250; H_3BO_3, 30; NaF, 8; and, KCl, 3 g/liter at pH 5.7 and I_c = amp/dm². This result is due to the fact that the microfissures which form in the plate through interaction of the periodic load and the internal stress function as stress concentrators on the steel surface and reduce the fatigue strength. Barklie and Davies point out that the tensile strength falls with increasing plate depth and internal stress.

It was also shown that compressive stresses arise in zinc plates laid down on steel from an electrolyte containing: $ZnSO_4$, 143; CH_3COONa, 0.25; and, gum arabic, 1 g/liter at pH 4.3, 18°, and I_c = 1.0 amp/dm².

Equally interesting results could be obtained from a study of fatigue strength in the case of deposits with high compressive stress. A particular instance of this kind is that of the deposition of copper in the presence of considerable quantities of thiourea.

No reduction in fatigue strength was observed in investigations of the effect of lead plates. These plates were essentially free of internal stress. Use of an intermediate layer of lead proved to be advantageous in the nickel plating of steel, no reduction in fatigue strength occurring in this case. This intermediate lead layer was deposited from an electrolyte containing: $[2PbCO_3 \cdot Pb(OH)_2]$—150, 50% HF—240, H_3BO_3—105; and, glue, 0.2 g/liter at 18° and a current density of 1.0 amp/dm².

Similar studies were carried out by Kudryavtsev and Ryabchenkov [2] who showed that chrome plating of carbon steels would also reduce the fatigue strength. Chrome plating was carried out from electrolytes of the following compositions (g/liter).

I.	CrO_3 — 227.7	H_2SO_4 — 3.37	Cr_2O_3 — 5.3	Fe — 5.84
II.	CrO_3 — 199.0	H_2SO_4 — 1.38	Cr_2O_3 — 2.5	
III.	CrO_3 — 207.7	H_2SO_4 — 2.06	Cr_2O_3 — 3.3	

The method of plating from these electrolytes was such as to give chrome plates free of cracks. The results of fatigue measurements are shown in Table 29.

It is seen from this table that the fatigue strength is considerably reduced by chrome plating, the reduction reaching 22% in certain cases. The maximum reduction was observed in specimens which carried thick plates. Kudryavtsev and Ryabchenkov note that the fatigue strength of chrome-plated steel is increased by annealing at 100°C for three hours, but remains less than that of the original unplated steel.

TABLE 29. Fatigue Limit of Chrome-Plated Specimens

Character of surface	Conditions of chrome plating and type of plate	Depth of chromium, mm	Treatment after plating	Fatigue limit			Composition of electrolyte
				kg/mm²	for smooth sample, %	for sample with notch, %	
Smooth	Without chrome plate	—	—	24.5	100	—	—
With notch, h = 1.4 mm	The same	—	—	14.8	—	100	—
Smooth	I_c = 32 amp/dm²; 60; shiny chrome	0.03	—	20.3	83	—	I
"	Shiny chrome	0.10	—	19.5	80	—	I
"	" "	0.03	Tempered: 100°, 3 hr	22.2	90	—	—
With notch	" "	0.03	—	13.8	—	93	—
Smooth	" "	0.10	Tempered: 100°, 3 hr	22.5	92	—	—
"	" "	0.10	Tempered: 250°, 2 hr	19.2	78	—	—
"	" "	0.03	The same	21.0	86	—	—
"	I_c = 25 amp/dm²; 70°; milky chrome	0.03	—	24.5	100	—	—
"	Milky chrome	0.03	Tempered: 100°, 3 hr	19.5	—	—	II
"	I_c = 35 amp/dm²; 80°; milky chrome	0.03	—	<20.0	<82	—	II
"	I_c = 22 amp/dm²; 70°; milky chrome	0.03	—	20.5	84	—	II
"	I_c = 25 amp/dm²; 70°; milky chrome	0.03	—	20.4	82	—	III

Increasing the temperature of tempering to 250° slightly reduces the fatigue limit of specimens carrying thick chrome plates, but is practically without effect on samples carrying thin plates.

These authors believe the reduction in fatigue strength to be due to the presence of high residual tensile stresses, which were originally set up in the deposit during plating.

The effect of the temperature of the electrolyte and the current density on the fatigue strength of chrome-plated machine parts has been investigated by Rykova [3]. Here, plating was carried out in an electrolyte containing 250 g/liter CrO_3 and 2.5 g/liter H_2SO_4. The author notes that the fatigue strength of steel can be reduced by chrome plating, the reduction amounting to 34-40% in the case of a plate having an internal stress of the order of 2700-3100 kg/cm² and 22-27% in the case of a plate with an internal stress of the order of 1200-1500 kg/cm².

Glikman, Fedot'ev and Chernova [4] have carried out similar studies on electrodeposits of chromium, while Kudryavtsev and Ryabchenkov have investigated nickel, copper, and zinc deposits.

Rykova has shown that a reversing current will lower the fatigue limit by 9-12%, at most. Moroz and Kudryatsev [5] have studied the effect of zinc deposits obtained from electrolytes of the following compositions (g-equiv/liter):

I. ZnO — 1.0	II. ZnO — 1	III. ZnO — 1	IV. ZnO — 1.0
NaCN — 3.0	NaCN — 2.5	NaCN — 1.5	NaCN—2—1.0
NaOH — 1.0	NaOH — 2.5	NaOH — 2.5	NaOH — 3.0

It was observed that zinc plating from electrolyte II would reduce the tensile strength of steel from the normal value, 177.6 kg/mm², to 160.3 kg/mm². The authors consider that this effect is due to the penetration of hydrogen into the metal (Table 30).

Moroz and Kudryatsev also note that an increase in cyanide content of the solution will increase the evolution and occlusion of hydrogen and thus lead to a marked deterioration in mechanical properties.

TABLE 30. The Effect of Cyanide Content on the Hydrogen and Fatigue Strength of 30KhGSA Steel

Condition of sample	Time of zinc plating, min	Hydrogen content of deposit, weight, %	Fatigue strength, kg/mm²
Original	—	0.00015	177.6
After zinc plating in electrolyte I	90	0.014	168.3
After zinc plating in electrolyte II	5	0.0002	173.8
	30	0.0003	173.5
	60	0.0003	165.8
	90	0.0005	160.3

The cited experimental material gives evidence of the existence of a close relationship between fatigue strength and internal stress. The reduction in fatigue strength due to plating varies with the internal stress in the plate, plates which are free of stress giving practically no reduction. Relations of this type are accounted for by the ability of the internal stresses to act as a supplmentary stress concentrators. It is clear that the character of the stress must have a significant effect on the fatigue strength, a tensile stress being detrimental in this respect and a compressive stress, beneficial.

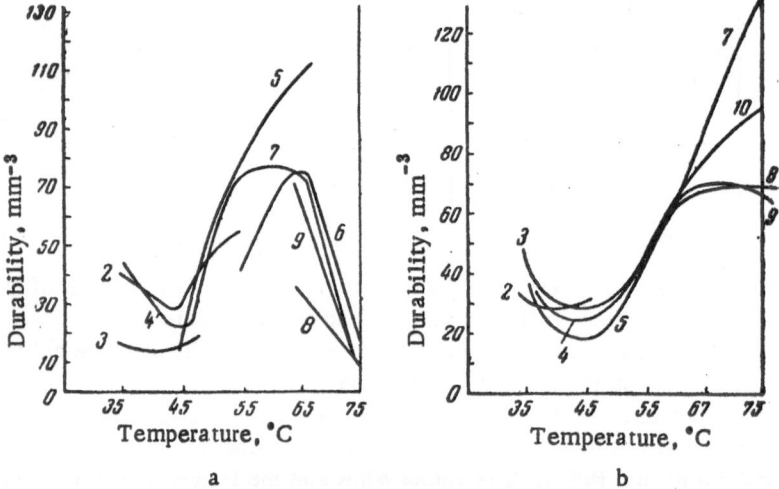

Fig. 101. The relation between the temperature of the electrolyte and the durability of chromium deposits (Shreider): a) Electrolyte containing 250 g/liter CrO₃; b) electrolyte containing 150 g/liter CrO₃. Current density: 1) 10; 2) 20; 3) 30; 4) 40; 5) 50; 6) 60; 7) 70; 8) 80; 9) 90; 10) 100 amp/dm².

The Durability of Chromium: Increase in durability is, in considerable degree, determined by the quality of the plate. Thus, a smooth chromium plate will increase the durability of cast iron by a factor of 6, while a chrome plate will give an increase of 30-150. The effect of various factors on the durability of chrome plates has been studied by many authors, but the results obtained are not entirely consistent. Nevertheless, it can be said that the resistance to wear depends on the quality and physicomechanical properties of the deposit and is, therefore, related to the composition of the solution and the conditions of electrolysis. Bogorad has studied the relation between durability and the conditions of electrolysis [6] and has found a minimum on the abrasion, current density curve.

Pertsovskii [7] has measured the durability of chromium obtained from a standard electrolyte at 49° and has shown that minimum abrasion corresponds to a current density of 62 amp/dm².

The experiments of Shreider [8] on the deposition of chromium from a standard electrolyte containing 2.5 g/liter of iron have shown that the durability, temperature curve has both a maximum and a minimum and remains essentially unchanged under alteration of the current density. It is interesting to note that minimum durability occurs at 45° (Fig. 101a). The form of the curve remains practically unchanged on passing to less concentrated electrolytes (CrO_3– 150, H_2SO_4– 1.5, CrO_3– 1.9 and Fe– 1.5 g/liter), minimum durability falling at 45° just as before (Fig. 101b). Increase in the current density increases the durability and displaces the maximum toward higher temperatures.

The durability hardness relation in chromium possesses a certain amount of interest. Figure 102 shows that the curve covering this relation passes through a maximum. This form of curve reflects the fact that loss on the left branch is essentially due to wear of soft metal, while loss on the right branch is due to grain chipping. It follows that the chromium deposits of maximum durability have a hardness of 650-925 kg/mm²

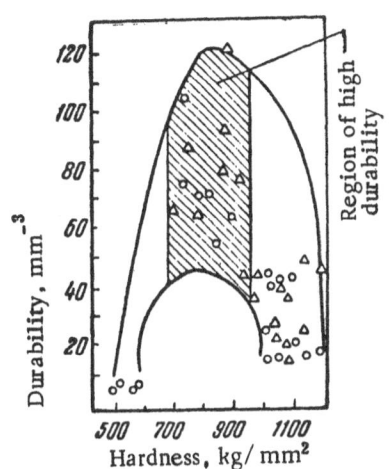

Fig. 102. The relation between hardness and durability in electrolytic chromium plates (Shreider).

Arkharova, Zagrubskii, Nemnonov [9] have expressed the opinion that durability is also related to the degree of textural completeness of the polycrystalline chromium deposit.

The Durability of Iron: The durability of electrolytic iron is less than that of chromium in a ratio of 1 : 3-1 : 5, but exceeds that of gray cast iron by a factor of 3 : 1 or more. Much data has been obtained on the

TABLE 31. The Durability of Iron and Its Alloys

Electrolyte	Ratio, durability of iron to durability of gray cast iron		
	method of mutual abrasion	abrasion on wall without lubrication	abrasion on wall with lubrication
Hot sulfuric acid, high-carbon iron electrolyte	2	6	4
Cold sulfuric acid, high-carbon iron electrolyte	3	9	6.8
Hot ferronickel electrolyte	1.4	2	—
Hot ferric sulfate electrolyte	2	5	—
The same electrolyte, deposit annealed at 650° for 1 hr	0.8	0.8	—

durability of iron and its alloys. Revyakin [10] has studied the durability of iron under various conditions and has concluded that it is possible to change the durability considerably by altering the conditions of electrolysis and the composition of the electrolyte (Table 31).

It should be noted that different methods of evaluating wear lead to the same general relations.

The table shows that the maximum durability is found in plates prepared from cold sulfuric acid, high-carbon iron electrolytes.

LITERATURE

1. R. H. D. Barklie and J. J. Davies, Engineer <u>150</u>, 670 (1930).
2. I. V. Kudryavtsev and A. V. Ryabchenkov, Methods of Surface Hardening of Machine Parts [in Russian] (Mashgiz, Moscow, 1940).
3. Collection, Studies on the Corrosion of Metals under Stress [in Russian] (Mashgiz, Moscow, 1953).
4. L. A. Glikman, N. P. Fedot'ev, and A. P. Chernova, Zavodskaya Lab. <u>17</u>, 1126 (1951).
5. I. I. Moroz and N. T. Kudryavtsev, The Elimination of Hydrogen Embrittlement of Steel in Electrolytic Zinc Plating [in Russian] (Mosk. Dom N.-T. Prop. im. F. E. Dzerzhinskii, 1957).
6. L. Ya. Bogorad, Korroziya i Bor'ba s Nei <u>5</u>, 379 (1937).
7. M. L. Pertsovskii, Porous Chrome Plating [in Russian] (Mashgiz, Moscow-Sverdlovsk, 1949).
8. A. V. Shreider, Collection, The Theory and Practice of Electrolytic Chrome Plating [in Russian] (Acad. Sci. USSR Press, Moscow, 1957) p. 77.
9. V. I. Arkharov, A. M. Zagrubskii, and S. A. Nemnonov, Vestn. Mashinostroeniya 10 (1940); S. A. Nemnonov, Tr. In-ta. Metallofiziki, Sverdlovsk, UFAN 2 (1944).
10. V. P. Revyakin, Izvest. Vyssh. Uch. Zav. Fizika 1, 132 (1958).